RAND NATIONAL SECURITY RESEARCH DIVISION

Internet Freedom Software and Illicit Activity

Supporting Human Rights Without Enabling Criminals

Sasha Romanosky, Martin C. Libicki, Zev Winkelman,
Olesya Tkacheva

Prepared for the U.S. Department of State

For more information on this publication, visit www.rand.org/t/rr1151

Library of Congress Cataloging-in-Publication Data
ISBN: 978-0-8330-9110-9

Published by the RAND Corporation, Santa Monica, Calif.
© Copyright 2015 RAND Corporation
RAND® is a registered trademark.

Support RAND
Make a tax-deductible charitable contribution at
www.rand.org/giving/contribute

www.rand.org

Preface

This research was conducted within the International Security and Defense Policy Center of the RAND National Security Research Division (NSRD) for the U.S. Department of State, Bureau of Democracy, Human Rights, and Labor, at the request of the U.S. Congress. NSRD conducts research and analysis on defense and national security topics for the U.S. and allied defense, foreign policy, homeland security, and intelligence communities and foundations and other non-governmental organizations that support defense and national security analysis.

For more information on the International Security and Defense Policy Center, see http://www.rand.org/nsrd/ndri/centers/isdp.html or contact the Director (contact information is provided on the web page).

Contents

Figures and Tables

Figures

Tables

Summary

The U.S. Department of State's Bureau of Democracy, Human Rights, and Labor (DRL) funds a major portfolio (henceforth, the "Portfolio") whose goal is to "promote fundamental freedoms, human rights, and the free flow of information online."[1] This goal is pursued through funding for various types of projects, each designed to "advance the rights and uphold the dignity of the most at risk and vulnerable or at-risk populations . . . [which] include women, lesbian, gay, bisexual and transgender (LGBT) individuals, religious and ethnic minorities, and people with disabilities."[2]

For example, DRL supports technologies that provide open access to information online and overcome censorship of information online, as well as secure communication technologies that help protect web and mobile messages from surveillance and eavesdropping. In addition, DRL supports training efforts and digital safety programs that provide relevant information and critical assistance for high-risk activists and those living in repressive countries. DRL also strives to support (1) research projects that evaluate the effectiveness of Internet freedom efforts and (2) advocacy efforts that empower civil society to inform national and international policymakers regarding the threats to free and open access of information, as well as potential mitigating solutions. Further, DRL strongly encourages the development of technolo-

---

[1]   See U.S. Department of State, "Bureau of Democracy, Human Rights and Labor Internet Freedom Annual Program Statement," web page, June 2, 2014.

[2]   U.S. Department of State, 2014.

gies that are available as open-source software and projects that foster strong partnerships with local organizations and human rights groups.[3]

In appropriating money to DRL (in addition to other government agencies that provide foreign assistance to Internet freedom), Congress has legislatively expressed its concerns that these projects may be used by criminals to further the commission of illicit activities and escape justice. For example,

> Provided further, That the circumvention technologies and programs supported by funds made available by this Act, shall undergo a review, to include an assessment of the protection against such technologies being used for illicit purposes . . . including an assessment of the results of these programs and safeguards against the use of circumvention technology for illicit or illegal purposes.[4]

DRL asked the RAND Corporation to examine the DRL Portfolio to determine the extent to which the projects it funds can be used for illicit purposes—specifically, whether DRL involvement has increased the potential for illicit use.[5] Note that this report does not explicitly evaluate the benefits of these projects in satisfying human rights objectives, although prior RAND work has done exactly that.[6] Information on each project was collected from a combination of publicly available information; in-person, telephone, and email conversations with grantees; and documents provided by DRL.

We applied a methodology designed to provide defensible and repeatable outcomes. First, we describe the technology or service of the project, and its benefit in promoting DRL's mission of fostering Inter-

---

[3]   U.S. Department of State, 2014.

[4]   U.S. House of Representatives, *House Report 112-331—Military Construction and Veterans Affairs and Related Agencies Appropriations Act, 2012*, 2012.

[5]   Throughout this document, we consider *illicit activity* to be synonymous with "criminal activity" (as would be considered by U.S. law).

[6]   See Ryan Henry, Stacie L. Pettyjohn, and Erin York, *Portfolio Assessment of Department of State Internet Freedom Program: An Annotated Briefing*, Santa Monica, Calif.: RAND Corporation, WR-1035-DOS, 2014.

net freedom across the world. Next, we apply the following three-part test as a means to consider the potential for the technology or service to be used for illicit purposes: Does it solve a criminal's communication problem? Does it provide a material advantage to criminals? Is the tool reasonably accessible to criminals? Finally, given the results of these tests, we assess whether DRL's involvement has increased the likelihood of the project to be used for illicit purposes.

Due to the sensitive nature of many of the initiatives being funded—and concerns regarding the safety of human rights activists who benefit from these projects—we omitted the actual names of projects, with all but one exception (Tor: The Onion Router). Instead, we examine groups of related services and technologies. The groups examined are digital safety, anti–distributed denial of service (DDoS), mesh networks, proxies/virtual private networks (VPNs), secure mobile communication, Tor, and a final category that describes two additional projects. Note that we provide an extended analysis of Tor due to its unique capabilities and widespread use.

From the analysis conducted, we conclude that the digital safety and anti-DDoS projects are not more likely to enable illicit activity because they either provide simple training material or require direct relationships with clients that would severely restrict any potential for illicit use of the services.

Mesh network projects provide mobile applications that enable low-bandwidth, ad-hoc network infrastructures over small geographic areas. While they could, in theory, be used to facilitate illicit activity, any criminal seeking strong anonymity or encryption would likely seek alternative technologies. There are also many other competing mesh network applications, so together these factors suggest that mesh network projects are not likely to be used for illicit purposes.

Proxy/VPN projects allow users to route their Internet traffic through an intermediary networked computer to bypass censorship and freely access public Internet services. While these technologies may encrypt the communication between the user and relay, they do not ensure strong anonymity, given that messages could be observable to the relay operator. VPN service provision is also at risk of being blocked by censoring states, which would further reduce the incentive for criminal actors to use these technologies. While the performance of VPNs in general makes them attractive for some criminals, the prevalence of non–U.S.-based VPN solutions suggests that these DRL-funded projects are not likely to be used for illicit purposes.

Secure mobile communication projects provide secure text, voice, and messaging capabilities. While confidentiality is provided by encrypting messages, anonymity of the sending device is not the primary feature. Any illicit activity would require additional technologies to ensure that a sender cannot be linked to his or her message (unlinkability). A number of alternative technologies also exist that provide comparable features and would therefore not make secure mobile communication projects more likely to be used for illicit purposes.

The Tor Project helps users circumvent censorship and detection through its distributed architecture and layered encryption. However, this protection is not absolute. Illicit users are still susceptible to mistakes and law enforcement investigative techniques. Moreover, the performance penalty incurred by the distributed architecture may be sufficiently great to deter high-bandwidth activity. Examination of available data suggests that a large portion of Tor traffic is conventional (unencrypted) web traffic, along with peer-to-peer (P2P) file-sharing traffic (some of which may be copyrighted files). Other data show that the most common destinations of Tor traffic are search, social media, and file-sharing websites, with U.S. users comprising only about 13 percent of all Tor requests. Localization, outreach programs, enabling human rights activities, and promoting transparency through metadata are just a few safeguards that Tor uses to ensure appropriate use of this technology. While the popularity and sophistication of this project could make it more likely to be used for illicit purposes, the capabilities provided by this technology existed prior to any DRL funding. We therefore conclude that DRL funding of Tor has not made this project more likely to be used for illicit purposes.

Finally, we examine two other projects funded under the DRL Portfolio: One provides a mobile and PC application useful for secure online storage, and the other is designed to provide Internet privacy from a bootable operating system on a USB flash drive. With each of these project groups, there exist numerous alternative solutions that would therefore not make these tools more likely to be used for illicit purposes.

In sum, this research finds that, in most cases, there is little reported evidence that the tools funded under this program assist illicit activities in a material way, vis-à-vis tools that predated or were developed beyond the Portfolio. Conversely, we conclude that they can and do provide crucial capabilities to *netizens* (non-criminal users of the Internet)—specifically human rights activists—either because they are freely available, easy to use, marketed and available only to human rights supporters, or because they operate in the user's native language. Further, given the wealth and diversity of other privacy, security, and social media tools and technologies, there exist numerous alternatives that would likely be more suitable for criminal activity, either because of reduced surveillance and law enforcement capabilities or fewer restrictions on their availability, or because they are custom built by criminals to suit their own needs.

Acknowledgments

The authors would like to thank Ryan Henry, Seth Jones, Stacie Pettyjohn, six anonymous law enforcement officers, an anonymous former Department of Justice attorney, an anonymous legal scholar, and many anonymous security, privacy, and Internet freedom experts for their valuable contributions and insights.

Abbreviations

DDoS   distributed denial of service

DRL   Bureau of Democracy, Human Rights, and Labor (U.S. Department of State)

FBI   Federal Bureau of Investigation

LGBT   lesbian, gay, bisexual, and transgender

NGO   non-governmental organization

P2P   peer to peer

Tor   The Onion Router

VPN   virtual private network

CHAPTER ONE

Introduction

The U.S. Department of State's Bureau of Democracy, Human Rights, and Labor (DRL) funds a major program (henceforth, the "Portfolio") to enhance Internet freedom. This Portfolio funds a set of tools (and associated services) that attempt to counter repressive governments' efforts to block online content or access to the public Internet, as well as these governments' attempts to prosecute individuals whose only crime is to exercise one of the basic freedoms specified in Article 19 of the United Nations' Declaration of Human Rights.[1] In appropriating funds to DRL, Congress legislatively expressed its concerns over how these projects would be used by asking for "a description of safeguards established by relevant agencies to ensure that programs are not used for illicit purposes."[2] The RAND Corporation was asked to examine this Portfolio to determine the extent to which the tools it funds could be used for illicit purposes and to recommend criteria by which future programs can be evaluated.[3]

Accordingly, we examined the benefits of these tools in promoting Internet freedom; their potential to be used for illicit purposes;

---

[1] Everyone has the right to freedom of opinion and expression; this right includes freedom to hold opinions without interference and to seek, receive, and impart information and ideas through any media and regardless of frontiers.

[2] U.S. Senate, *Senate Report 113-81—Department of State, Foreign Operations, and Related Programs Appropriations Bill, 2014*, 2014.

[3] Our assessment is limited to Internet freedom, rather than broader U.S. values such as social rights and anti-discrimination.

and, where possible, evidence of actual illicit use.[4] In doing so, we took into account the extent to which tools that preceded or evolved apart from the DRL program could support such illicit purposes. Finally, we examined possible safeguards that might discourage the use of the technologies for illicit purposes.

---

[4] While understanding the human rights context of these tools is useful for appreciating their intended purpose, this is orthogonal to an analysis of the potential illicit use of such tools.

CHAPTER TWO

Why Internet Freedom Tools?

Internet freedom efforts aim to promote democracy and human rights interests across the world by ensuring safe access to the global Internet. They proceed through various initiatives, including education, training, awareness campaigns, software, and information technologies, that together foster free and open access to the public Internet— free of surveillance, censorship, and harmful repercussions by repressive governments. For example, some of these commonly used software applications help ensure the privacy and safety of *netizens* (used here to refer to all non-criminal users of the Internet) by masking the origin of their communication, routing it through intermediary servers across the globe, and encrypting their messages.

Many people benefit from the privacy and security capabilities provided by the tools discussed in this report.

Dissidents and human rights activists, for instance, use these tools to communicate information about the atrocities and oppressive behaviors they witness. Journalists use these technologies to upload videos they have taken of human rights violations or war crimes. Individuals in highly oppressive regimes use these tools to circumvent government censorship to gain access to the public Internet.[1]

---

[1] For example, Gambia recently passed a law imposing criminal fines of $100,000 and 15 years in prison for individuals who use the Internet to "spread false news against the government, incite dissatisfaction or instigate violence against the government, caricature, abuse or make derogatory statements against public officials." See Modou S. Joof, "'Internet Is Being Used as a Platform for Nefarious and Satanic Purposes,'" Front Page International, July 28, 2013.

Others use these tools simply to prevent disclosure of their digital identity, which may be leaked or tracked as they browse online. These individuals have nothing to hide and are not concealing any illicit activity—they simply have strong privacy sensitivities and prefer to not be tracked as they access news, social networks, and e-commerce services on the web.

Academics and news organizations use the tools to conduct research and interviews and to anonymously transfer documents containing information about allegedly corrupt corporate or government practices. Religious, LGBT (lesbian, gay, bisexual, and transgender), racial, and ethnic minorities also use these privacy technologies to protect their safety and associate freely and anonymously, justifiably fearing that identifying themselves may lead to harm.

Law enforcement officers from the United States and across the globe use these tools to covertly identify, track, and apprehend criminals. The anonymity features of these tools allow agents to conceal the origin of their communications when connecting from government networks.

Finally, U.S. and international non-governmental organizations (NGOs) train and support many international groups who, in turn, train local citizens on safe and appropriate use of the Internet. For example, one U.S.-based NGO provides secure laptop computers to local groups in foreign countries with information about encrypted email, disk encryption, and strong password hygiene tools to protect users' digital and physical safety.

Empirical evidence confirms surges in the use of Internet freedom technologies following dramatic events such as revolutions. In some cases, the data also reveal an equally dramatic decline in Internet activity caused by swiftly imposed government controls. For example, data reported from one popular anti-censorship tool saw daily user activity (as measured by web-based user requests) in Tunisia rise dramatically—from five million hits to more than 30 million hits—in January 2011, as citizen activists and journalists shared their messages across the world. Similarly, daily activity by Libyan users rose from a few hundred thousand to almost 30 million hits between mid-February and March 2011 (during the revolt against Colonel Muammar Qaddafi) before being abruptly cut off. The Egyptian revolution saw a similar surge—from a few million to more than 60 million daily hits. Further examples can be drawn from Vietnam, Pakistan, and China.

Surges in Tor (The Onion Router) usage activity from 2011 in Egypt and Libya are shown in Figure 2.1. The left panel shows Tor client usage in Egypt increasing from 500 daily users to more than 2,000 in early 2011. The increase coincided with the January 25 revolution, which resulted in the overthrow of President Hosni Mubarak.[2] The right panel of Figure 1 shows Tor usage in Libya, which increased from around 50 daily users to almost 300 in early 2011. This surge coincided with the Libyan civil war, which began on February 16, 2011, and ended in the death of Colonel Muammar Gaddafi.[3] Both panels also show the subsequent drop in usage in the face of government censorship.

Although Internet freedom projects were designed, developed, and distributed to promote human rights and freedoms across the globe, they can also be abused to conduct illicit activity—just as with all technologies. The same technology that allows ambulances to speed to hospitals and save lives also allows bank robbers to power their getaway vehicles. The same technology that lets dissidents communicate with one another to explore the boundaries of freedom in authoritarian states also allows those engaged in illicit activity to communicate with one another, for example, to exchange drugs in democratic, law-abiding states. So it is with DRL-funded technologies and services.

For example, intelligence operatives operating under the authority of foreign governments use these same security and privacy techniques to carry out reconnaissance and counterintelligence missions seeking information on economic sanctions and antinuclear proliferation.[4] Operatives are usually well funded and highly sophisticated in

---

[2] "Timeline: Egypt's Revolution," *Al Jazeera*, February 14, 2011.

[3] "Libyan Uprising One-Year Anniversary: Timeline," *The Telegraph*, February 17, 2012.

[4] Siobhan Gorman, "Iran-Based Cyberspies Targeting U.S. Officials, Report Alleges," *Wall Street Journal*, May 29, 2014.

Figure 2.1
Estimated Tor Usage in Egypt and Libya, 2011

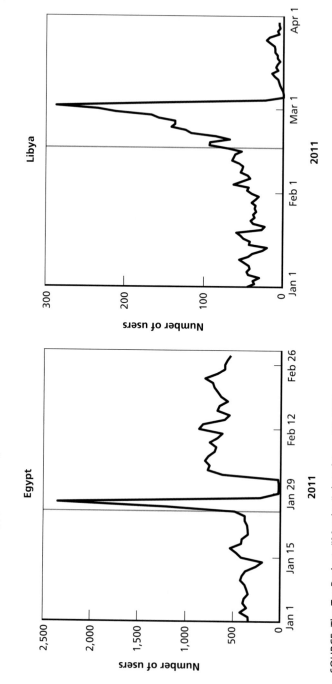

SOURCE: The Tor Project, "Metrics," dated June 30, 2014.
NOTE: The vertical line in the panel for Egypt marks January 25. In the panel for Libya, the line marks February 16.
RAND RR1151-2.1

their use of surveillance and espionage tools and techniques, as well as in their ability to conceal their activities.[5]

Terrorists use these tools because they are willing to sacrifice performance (speed of communication) to gain additional secrecy and security of communication. Over the past decade, terrorists groups have become much more sophisticated in their use of encryption. While those who plotted the September 11, 2001, attack did not even encrypt their communications,[6] nowadays al Qaeda builds its own privacy-protection tools because it does not trust those developed by Western companies.[7]

Child pornographers employ many kinds of software tools to transmit and view illicit images covertly. Since mere possession of child pornography is a criminal offense, the secrecy and anonymity both of its communication and of the tools employed are of great importance. Name recognition or online reputation is not important due to a high rate of turnover of child pornography sites. Instead of using web addresses or site names, the search is conducted using keywords that serve as code words in pedophile communities.[8]

Members of organized crime groups may employ sophisticated digital methods to conceal their activities even as they also use traditional and non-technical forms of communication such as physical couriers or dead drops.

Thieves who steal personal information for financial gain may have only simple requirements for concealing their transactions, using obfuscated instant messaging, encrypted virtual private networks (VPNs), or even publicly accessible social networking tools.

---

[5] For example, see recent stories relating to China, Iran, and Russia: Shane Harris, "Exclusive: Inside the FBI's Fight Against Chinese Cyber-Espionage," *Foreign Policy*, May 27, 2014.

[6] Emil Protalinski, "Osama bin Laden Didn't Use Encryption: 17 Documents Released," blog post at *ZDNet.com* website, May 3, 2012.

[7] Recorded Future, "How Al-Qaeda Uses Encryption Post-Snowden (Part 1)," May 8, 2014a; Recorded Future, "How Al-Qaeda Uses Encryption Post-Snowden (Part 2)—New Analysis in Collaboration with ReversingLabs," August 1, 2014b.

[8] Patrick Forde and Andrew Patterson, "Paedophile Internet Activity," *Australian Institute of Criminology: Trends and Issues in Crime and Criminal Justice*, Vol. 97, November 1998.

CHAPTER THREE

Internet Freedom Tools Are Countermeasures to Repression

Years ago, it was hoped that the Internet would become a communications technology that would allow the world's netizens to access global information that was previously available only in one-to-many mass media (in contrast to social media, which have far more one-to-one messaging). Such media could be dominated by state agencies (e.g., television) or physically controlled (e.g., by confiscating printed material). According to early Internet enthusiasts, cyberspace would be a domain free from government interference, where any restriction was tantamount to damage that the Internet routinely routed around. An era of freedom and global understanding beckoned.

Although repressive governments of the world did not immediately respond to the fillip that the Internet gave to freedom, they have now done so and in a variety of ways:

- In some cases (e.g., Cuba, North Korea) governments limit Internet access, either absolutely or in effect, by imposing costs and restrictions for (nearly) everyone. Other countries limit Internet access at specific times (e.g., in Egypt during the Arab Spring) or places (e.g., where people gather to exercise their freedom to assemble). Both Jordan and Russia have also recently been increasing efforts to restrict Internet access.[1]

---

[1] Sanja Kelly, Mai Truong, Madeline Earp, Laura Reed, Adrian Shahbaz, and Ashley Greco-Stoner, eds., *Freedom on the Net 2013: A Global Assessment of Internet and Digital Media*, Freedom House, October 3, 2013.

- More sophisticated countries (e.g., China, Iran) institute selective blocks on specific sites (e.g., Google News) and online content (e.g., any mention of "Tiananmen Square"). Such countries also target prominent sites and prevent anyone from accessing them by using distributed denial-of-service (DDoS) attacks.
- Some governments (e.g., Belarus, Vietnam, Bahrain) hijack Internet services and technologies to identify dissidents and political opponents, either overtly (e.g., by tracing Facebook "friendings") or covertly (e.g., by infecting dissident computers with beaconing malware).[2]
- Many governments (e.g., Turkey, Bangladesh, Azerbaijan) criminalize the expression of dissidence and use the Internet as a tool of surveillance.[3]

The converse of these government-imposed measures provides a first-order approximation of an Internet freedom agenda: countermeasures that would enhance Internet availability (at all places and times), circumvent censorship, counter DDoS, improve personal computer security, and/or increase the anonymity of communications.

Each of these countermeasures corresponds and responds to measures that governments impose on those seeking to exercise their human rights. If these repressive measures had not been imposed in the first place, most of the Internet freedom agenda would not be necessary.[4]

Most of these repressive government measures were not meant to suppress (what the United States would consider) illicit activities. Governments, one would imagine, do not restrict Internet use for everyday citizens for the primary purpose of reducing, say, street crime. Similarly, although a government may block sites used to support what it considers (not entirely unreasonably) criminal activity (e.g., adult por-

---

[2] Kelly et al., 2013.

[3] Kelly et al., 2013.

[4] Some Internet freedom technologies protect the rights of sensitive minorities (e.g., from the LGBT community) with legitimate social reasons for preferring anonymity—even in the absence of specific government measures against them (e.g., statutes that criminalize certain sexual acts).

nography), the sites that support what the United States would consider criminal activity are not the main focus of their censorship. Similarly, DDoS is not the most common government approach to the suppression of criminal activity. Likewise, use of malware by repressive governments to find criminal activity in Western countries, while it does happen, does not take place on a routine basis. The only measure available to governments that could be as great a threat to criminals as to, say, human rights activists is the interception of communications to collect evidence of crimes.

This should be kept in mind when differentiating between the use of DRL tools to enhance Internet freedom and the subversion of these tools to support illicit activity. To wit, of the five countermeasures represented by DRL tools (some of which encompass more than one countermeasure), the most likely place to look for criminal activity would be among those tools that hide the identity of users and the information they exchange.

CHAPTER FOUR

How Could DRL Funding Affect Criminal and Netizen Behaviors?

We will now explore a theoretical framework that examines differences between two types of users of Internet freedom tools: netizens and criminals. Both groups seek to communicate safely to protect their identities and avoid any harmful consequences.

Many Internet freedom tools provide a variety of capabilities, such as circumvention, encryption, and anonymity. Further, some of these tools are offered for free, while others are available for a cost; some of the tools are open-source while others are not; some are available legally, while others are only obtainable through illegal channels. Because DRL is not the only supplier[1] of Internet freedom tools, we can separate such tools into those funded and those not funded by DRL. We further assume that both netizens and criminals enjoy utility from using a mixture of both sets of tools.[2]

However, each group experiences constraints on their free and unfettered use of these tools. Netizens are often limited in their access and choice of tools, either because of cost or the lack of training, education, or availability. For criminals, DRL-funded tools include safeguards and design features that may restrict their use for illicit purposes. For instance, in many cases, the distribution and availability of

---

[1] We use the term *supplier* in the familiar economic sense of a producer of a normal good. DRL does not actually "supply" tools, but rather funds their development.

[2] The specific composition of services and tools is not critical for this discussion, but each user type exhibits some preference and enjoys utility from the use (consumption) of these tools.

the service or technology is provided only to those personally known to program participants.

We first examine the trade-offs made by each group of individuals when selecting among DRL-funded and non–DRL-funded tools, and identify the effect of DRL funding on each group.

Figure 4.1 illustrates the change in utility experienced by netizens before (left panel) and after (right panel) DRL funding. The x-axes show the level of consumption of non–DRL-funded tools, while the y-axes show the level of consumption of DRL-funded tools.[3]

First, examine the left panel of Figure 4.1. The curved line (U_1) illustrates the maximum utility (benefit) enjoyed by a notional netizen when selecting a mixture of non–DRL-funded tools (x-axis) and (to be) DRL-funded tools (y-axis).[4] The further outward from the origin along a 45-degree angle, the greater the utility enjoyed. But this netizen's maximum utility is limited by the practical challenges she faces in accessing, affording, and using any of the Internet freedom tools. This constraint is illustrated by the straight line, C_1 (i.e., the constraint

Figure 4.1
Effect of DRL Funding on a Netizen

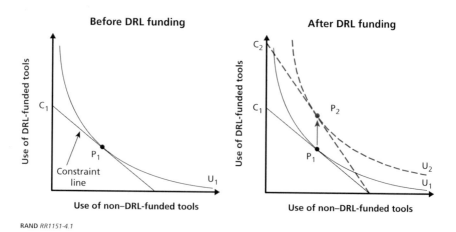

---

[3] We make the implicit assumption that a DRL-funded tool would have existed without DRL funding.

[4] Formally, this is an indifference curve showing the consumption of two normal goods.

line). The tangency of these curves (point P_1) identifies the maximum possible utility enjoyed, as well as the relative amounts of tools used.

In the right panel, we show the key effects of DRL funding for this netizen. The funding of Internet freedom tools enables better access to technologies and services, allowing her to, among other benefits, freely access information resources and communicate with others in a manner that preserves her personal and digital safety. In effect, DRL funding relaxes the constraints imposed on her (which is tantamount to lowering the price of using such tools), allowing her to consume more DRL-funded tools and enjoy more total utility. In our model, DRL funding increases the accessibility of DRL-funded tools, thereby changing the constraint from line C_1 to line C_2, as shown.[5] This improvement in available resources allows her to enjoy an increase in utility (point P_1 to P_2), as shown in the change from curve U_1 to U_2.[6]

Next, we examine the effects of DRL funding on criminal interests, as illustrated in Figure 4.2.

Similar to what is shown in Figure 4.1, the left panel of Figure 4.2 illustrates a notional utility function of a criminal actor before DRL funding. The curved line, U_1, identifies the indifference between tools as a criminal employs alternative combinations of Internet freedom tools (which, in these circumstances, are used to commit crimes). Like all users, criminals also face constraints on their time, budget, and availability to acquire and use these technologies and services (identified by the straight constraint line, C_1). Therefore, she obtains maximum utility at the tangency of these two curves, shown by point P_1.

The right panel illustrates the changes to consumption of DRL-funded tools, and the subsequent reduction in overall utility. First, while the funding of the tools on the y-axis may provide an initial benefit to the criminal (just as it does with netizens), the existence of safe-

---

[5] We assume that funding DRL tools has no effect on the properties of non–DRL-funded tools.

[6] Note that, strictly speaking, DRL funding reduces the cost of adoption of DRL-funded tools relative to non–DRL-funded tools. This change may have multiple effects on the constraint curve, including to both pivot and shift it. Strictly speaking, these are referred to as *income* and *substitution* effects; however, for the purpose of this discussion, we simplify the analysis to simply show the repositioning of the constraint curve.

Figure 4.2
Effect of DRL Funding on Criminal Actors

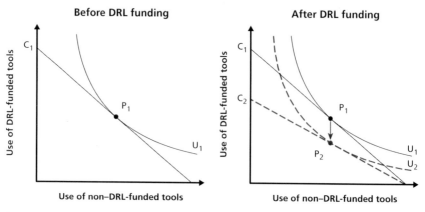

guards and design features serves to limit and restrict the use of these services and tools by criminals. In effect, this changes the constraint faced by a criminal from line C_1 to line C_2. Given this new constraint, the criminal's overall utility reduces from P_1 to P_2 as a result of such safeguards.

These figures illustrate how the funding of Internet freedom tools (with appropriate safeguards) can improve the benefit to netizens without necessarily increasing (and even possibly decreasing) their use by criminals. Further, one can argue that, in the presence of numerous alternative security and privacy tools, criminal incentives to use DRL-funded tools would be weak, because their advantage in using them rather than other tools would be similarly weak.

CHAPTER FIVE

Do Netizens and Criminals Seek the Same Things from Internet Freedom Tools?

The next step toward understanding the uses and abuses of Internet freedom tools begins by identifying differences in needs between netizens and criminals. This chapter, therefore, examines potential differences in preferences for technological features of Internet freedom tools across these two groups. To do so, we first examine surveys of bloggers.

Netizens turn to Internet freedom tools, in large part, to access blocked sites, such as Facebook and Twitter, to post politically sensitive content for a broad audience.[1] When bloggers worldwide were asked about the relative importance of privacy versus circumvention features of the tool, the percentage of those who used Internet freedom tools for circumvention (54 percent) was higher than those who used them for privacy protection (45 percent).[2] Bloggers, after all, face a trade-off between cultivating name recognition and protecting their identity. Since the credibility of blog posts and tweets depends on the audience's familiarity with the source, the probability that anonymous posts will resonate with the wide audience is much lower, unless the posts are reposted on a reputable news outlet.

Since there are multiple tools for accessing blocked content and protecting privacy, it is important to understand how netizens choose among alternative Internet freedom tools. Survey evidence suggests that there might be differences in preferences across countries and user

---

[1] Robert Faris, John Palfrey, Ethan Zuckerman, Hal Roberts, and Jillian York, *International Bloggers and Internet Control: Full Survey Results*, Cambridge, Mass.: Harvard University Berkman Center for Internet and Society, August 18, 2011.

[2] Faris et al., 2011.

types. For instance, international bloggers' ranking of the importance of Internet freedom tool features differed from the ranking provided by netizens in China. Bloggers identified "privacy," "ease of discovery and downloading," and "ease of installation" as the three most important categories.[3] Chinese netizens (the majority of whom were students) listed "reliability," "speed of connection," and "ease of installation" as the most important categories. Chinese netizens were also attracted to Internet freedom tools that were built into commercial platforms— Google Tools and Amazon—because the economic costs of blocking access to those platforms were so prohibitive that the Chinese authorities were reluctant to block them.[4]

Several important takeaways emerge from these surveys. Since most content that netizens seek to access is in the public domain (i.e., the website addresses are known), netizens do not need to coordinate their choice of Internet freedom tools with either website servers or with other netizens who also seek to access those websites. Communication among netizens also takes place in the public domain and targets a wide audience. As one Saudi activist observed, Twitter enabled him to turn "kitchen" conversations that he had with a small circle of friends into a public discourse. Although many participants in these public conversations would like to have their identities protected to avoid sanctions by the authorities, the content of their discussion is intended for a broad audience.

The rapid availability of Internet freedom tools also affects their utility for netizen use. Political events unfold very rapidly, and the timing of when information is accessed and shared affects its impact by changing the nature of public discourse. When netizens are able to communicate their story before that offered by state-controlled media, they become the agenda-setters. This forces the regime to play a catch-

---

[3] Faris et al., 2011, p. 31.

[4] David Robinson, Harlan Yu, and Anne An, *Collateral Freedom: A Snapshot of Chinese Internet Users Circumventing Censorship*, OpenITP, April 2013, p. 11.

up game by offering an alternative account, rather than *the* account of the event.[5]

In most cases of illicit activity, by contrast, the information shared online has the property of a *club good*. Unlike the netizen, who seeks to distribute information into the public domain, the criminal seeks to restrict access to information to a small subset of network participants or Internet users because, in most cases, it increases the material gain from pursuing an illicit activity. For example, a distributor of child pornography can increase his or her material gain by restricting access to the illicit content by granting access only to those members who either pay monthly dues (via anonymous currency) or share their libraries with other users. Drug dealers might want to limit access to information about the pick-up locations only to carefully vetted online users. Thus, rather than seeking to increase the size of the audience that receives a specific message, more often than not, actors who turn to the Internet for illicit activity seek to reach a narrow and a carefully selected audience.[6]

In addition, to avoid detection by law enforcement officials, suppliers of illicit content must frequently change the Internet address of their website, making cultivation of name recognition both pointless and infeasible. The suppliers of illicit content then turn to focal points (i.e., the Silk Road network) to increase the probability of a transaction. Instant access is far less important because the content is not affected by the political context.

In further contrast, the security of communication among criminals is only as good as its weakest link. Failure to encrypt communication by one group member can jeopardize the security of communication. Similarly, for an online network to emerge as a focal point it should be supported by Internet freedom tools that are widely used to increase the number of potential new club members. This frequency-induced property of technology creates a lock-in effect that makes uni-

---

[5] This is assuming that netizens will provide a true account, which may not be the case, either on purpose or by accident, or even possible, due to the subjective experiences of all participants.

[6] Terrorist online intimidation campaigns are a rare exception to this pattern.

lateral shifts from one technology to another not utility-enhancing. For some types of illicit activity, a critical mass of technology users might be required before the migration from one type of Internet freedom tool to another can happen.

Overall, therefore, we find that time sensitivity, reputation building, and the public nature of the netizens' discourse are key characteristics of their online activity. On the other hand, online criminal behavior exhibits a typical club good, is less time sensitive, and may suffer from higher switching costs.

CHAPTER SIX

Methodology

Having examined the different needs of netizens and criminals, we now offer the following methodology, which was designed to provide defensible and repeatable outcomes.

We first provide a brief description of the specific technology or service (hereafter "tool" or "project").[1] Next, we identify the intended purpose of the project, as described by its implementers. That is, we describe its benefits in promoting DRL's mission of Internet freedom across the world. We then examine the extent to which these tools can be used for illicit purposes.[2] Our purpose is not to prove that a given tool could never be used for criminal activity, but to understand the potential that such tools offer for criminal use.

To address this challenge, we proposed a minimum set of criteria that must be satisfied for a project to be used for illicit purposes: First, it must address a specific problem of an illicit actor; second, it must provide a materially better capability than tools developed prior to, or independent of, DRL's efforts; and, third, it must be reasonably available to the criminal and free of significant safeguards. In order for DRL involvement to produce an affirmative answer, the tool (or project

---

[1] Note that because we examine groups of technologies and services, there may be individual variation across the dimensions that compose this analysis. In cases where there is a material difference in behavior or capability, we identify the capability, but without uniquely identifying the specific technology platform.

[2] Note that because our challenge is to examine how these projects could support criminal activity, we describe, but do not systematically evaluate, their benefits to netizens.

or service) must meet each of the three criteria. We next examine each of these conditions in turn.

Regarding the first criterion, that the technology or service must address a substantial digital communication problem for a criminal, we assume that a project that solves no such problem would be of little interest to criminals and therefore fails this test.[3] Following a simplified taxonomy identified by Gambetta (2011),[4] we consider that criminals face three common problems. First, criminals seek to communicate secretly with other *known* colleagues without being surveilled, such as when planning and executing a criminal activity. This is termed the *communication* problem. Second, criminals seek to identify and communicate with others whom they have never met—e.g., to recruit new members, raise funds. This is termed the *identification* problem. The third problem is the challenge of selling stolen goods or offering illegal services. This is termed the *advertising* problem. Therefore, we argue that *illicit activity using a DRL-funded tool is more likely when the tool solves at least one communication, identification, or advertising problem.*

The second criterion, which is conditional on the DRL-funded tool satisfying the first test, is that the tool must also provide a capability to criminals that is materially better than any other tool not funded by DRL or that would have existed absent DRL funding.[5] We therefore also examine whether alternative tools—those not funded by DRL—could be used for the same purpose. For example, compare a software application that could guarantee absolute privacy forever[6] with a short

---

[3] Here, we recognize that it is not practical or necessary to consider every single difficulty that each criminal type would encounter. Instead, we believe it is sufficient to consider general matters that most criminals and criminal groups would reasonably face during the course of their activity.

[4] Diego Gambetta, *Codes of the Underworld: How Criminals Communicate*, Princeton, N.J.: Princeton University Press, 2011. While Gambetta's taxonomy refers generally to the act of signaling messages and intentions as well as direct communication between parties, there are useful applications when examining the communication needs of both human rights and criminal groups.

[5] Our assumption of a material capability is one that would provide a substantial or fundamental improvement that would not otherwise be available.

[6] Chosen for illustrative purposes only, as this is clearly an unobtainable property.

guide on protecting e-mails. The former capability would substantially help a criminal evade law enforcement (which would therefore pass this test), while the latter could easily be obtained elsewhere. Where available, we also examine any evidence of actual illicit use. Therefore, we argue that *illicit activity using a DRL-funded tool is more likely when it provides a material capability to criminals that would not have otherwise existed without DRL funding.*

Finally, even if the tool does solve a problem and is materially better, the criminal must still be able to use it, either because he or she has reasonable access to it or because there are few safeguards preventing its use. For example, those tools or services that are only available to a select group of vetted individuals would be unlikely to be used by criminals and therefore fail this test. Similarly, those tools that include sufficient safeguards or other design characteristics that would restrict their use would also provide little benefit to criminals. On the other hand, tools that are publicly available would pass this test. Therefore, we argue that *illicit activity using a DRL-funded tool is more likely in the absence of any safeguards that would reasonably prevent a criminal from obtaining or using it.*

CHAPTER SEVEN

Analysis of DRL Internet Freedom Projects

While each project has been examined individually, it is the policy of DRL not to reveal the identity of grantees without explicit consent. Further, RAND was asked to evaluate collections of projects, categorized according to the capabilities they afford.[1] Therefore, we evaluated the DRL projects in terms of the following groupings: digital safety, anti-DDoS, mesh networks, proxies/VPNs, secure mobile communications, Tor, and two projects grouped into a final category called "other." Note that we uniquely identify and provide an extended analysis of the Tor project, because of its popularity and sophisticated capabilities.[2]

Digital Safety

Digital safety projects cater to disadvantaged, vulnerable, or isolated peoples, or those living in countries that impose censorship on their media. These projects support Internet freedom by educating and training netizens on the risks of online communication, in addition to providing adaptation and localization (i.e., language translation) services for already existing circumvention and secure communication tools.

---

[1] While there is some variation across the individual tools and services within these groups that will unavoidably be lost in the overall description, no variation resulted in qualifying our conclusions.

[2] Consent was provided by the Tor Project to identify this project uniquely.

Other forms of in-person training are provided to at-risk journalists, technology activists, human rights workers, and bloggers, and include both technical discussion and practical training regarding physical safety. Delivery of training material may include presentations, group discussions, individual tutoring, and hands-on training. Examples include instructions and tasks designed to teach the use of specific security and privacy tools, demonstrations of wireless network attacks, tips for avoiding network surveillance, and proper protection of one's mobile devices and home computers. The training itself is often provided in whatever physical locations and over whatever Internet connections are available.

In addition to these training programs, other projects aggregate news and various forms of localized content (i.e., content local to particular regions and countries) and ensure distribution to people who would otherwise be unable to receive it. In addition, many of these projects focus specifically on particular countries or regions that are especially impacted by state-sponsored censorship.

We now apply the three-part test for illicit use of digital safety tools. Because these training efforts are generally conducted in small venues, person-to-person, there is little ability for them to satisfy any of the communication, identification, or advertising problems faced by criminals. Further, because the training projects often concern basic user privacy and Internet safety, they would not provide any material advantage to criminal actors. Third, the grantees supporting these projects often work closely, even in person, with local NGOs to build and maintain strong relationships of trust with local citizens and activists across the world. This implementation strategy provides a strong safeguard against illicit use of any training or education. The grantees also foster strong relationships with local non-profit groups to ensure that they operate appropriately in each country.

Therefore, because these digital safety projects do not satisfy any of the conditions of the three-part test, we conclude that they are not more likely to be used for illicit purposes, relative to non-DRL solutions.

Anti-DDoS

The anti-DDoS projects supported by DRL help non-profit and civil-society organizations maintain an online presence by ensuring resilience against DDoS attacks.[3] These services protect a client's website by using Internet blacklists and basic firewalling techniques that filter and absorb malicious traffic. In addition, they employ other technical capabilities, such as load balancing, short-lived domain name system resolution, and reverse proxy caching. These anti-DDoS services are subscription-based and are specifically offered at low cost or for free to non-profit groups that may be the target of oppressive actors because of their human rights–based work.

Next, we apply the three-part test for illicit use. In regard to solving problems of illicit users, anti-DDoS services would be most suited to addressing the advertising problem because they help provide access to a publicly available website. Next, the capabilities provided by these services are not unlike those for commercially available tools, and, in some cases, provide a much broader set of traffic management services and cyberattack protection at a higher cost—for example, CloudFlare (www.cloudflare.com), Rackspace (www.rackspace.com), and Amazon Web Services (aws.amazon.com). Third, these services are operated by the project owners themselves, rather than by the users. Therefore, to enjoy protection, organizations must comply with and uphold a set of ethical requirements to ensure that they are, indeed, supporting human rights and Internet freedom. This vetting of potential clients provides a strong safeguard against illicit use of these anti-DDoS services.

Therefore, these services only partially satisfy the first test (the advertising problem), but fail the second test (because there exist alternative services that are more readily available to illicit users) and the third test (because recipients of anti-DDoS services are vetted for compliance with human rights efforts). We conclude that the anti-DDoS services funded by the DRL Portfolio are not more likely to be used for illicit purposes, relative to non-DRL solutions.

---

[3] DDoS attacks are those that seek to deny access to an Internet service (e.g., a website) by overloading the computer systems with activity.

Mesh Networks

Mesh network technologies provide a communication infrastructure that operates separately and independently from existing wireless or cellular communication channels, and are therefore most useful in locations where reliable or private access is unexpectedly unavailable (e.g., in the face of temporarily high demand), heavily censored, or subject to repressive surveillance. Mesh networks that use mobile phones are designed for ad-hoc, spontaneous, and temporary gatherings of people, such as during protests or community events. Some mesh networks operate by establishing a peer-to-peer (P2P) network among personal computers or mobile devices whose users are separated by dozens or hundreds of meters. By stringing enough links together, it is possible to provide connectivity to a public network from just one device. For instance, many users within a city block might relay their messages through the Internet connection of just one individual. Larger base stations or antennas can also be used to extend the effective range of these private networks.

Mesh networks are valuable in cases of citizen protest, where an oppressive government suddenly disables public Internet access (as illustrated, for instance, in Figure 2.1). In such cases, participants form an ad-hoc network to share information and coordinate activities. In addition, mesh networks become critical out-of-band communications systems during natural disasters, so as to help organize and distribute humanitarian need. Indeed, as one journalist writes, "[mesh networks] provide a means for people to self-organize into communities and share resources amongst themselves: Mesh networks are operated *by the community, for the community*."[4] Further, "with mesh networking, people are building a community-grown network infrastructure: a distributed mesh of local but interconnected networks, operated by a variety of grassroots communities. Their goal is to provide a more resilient system

---

[4] Primavera De Filippi, "It's Time to Take Mesh Networks Seriously (and Not Just for the Reasons You Think)," *Wired.com*, January 2, 2014 (emphasis in original).

of communication while also promoting a more democratic access to the internet."[5]

We now apply the three-part test for illicit use of mesh networks. First, because mesh networks are designed to provide network access for areas with unreliable (or blocked or censored) communication infrastructure, they would, at most, address only the communication problem of criminals who seek to meet in a small geographic location to plan or execute an activity.[6] These services would not be useful, for example, for solving advertising problems because they do not directly facilitate advertising of illicit services, especially beyond the limited range of mesh networks (several hundred meters).

Second, the specific projects within the DRL Portfolio are generally in the early stages of development and therefore would not provide any material capabilities to criminals that would not exist elsewhere. An example of a non–DRL-funded tool is Edge Velocity,[7] a for-profit corporation that provides mesh networking solutions for emergency responders. The Athens Wireless Metropolitan Network, the Guifi network in Spain, and the Free Network Foundation in Kansas are other commercial examples of non–DRL-funded mesh networking tools.[8] The FireChat iPhone application has seen significant recent usage in Iraq and Hong Kong.[9]

Third, the mesh network technologies funded under the DRL Portfolio have been chosen, in part, because they are open-source and can be made freely available to cater to human rights activists across the world. While it is conceivable that mesh networks could be employed

---

[5] De Filippi, 2014.

[6] Of course, mesh networks also help solve the communication and identification problems for activists seeking to communicate in rallies and protests, and to warn one another (even total strangers) of any danger.

[7] Edge Velocity Corporation, "About Us," web page, undated.

[8] Clive Thompson, "How to Keep the NSA Out of Your Computer," *Mother Jones*, September–October 2013.

[9] Russell Brandom, "Iraqis Seek Out New Tools to Blast Through Internet Blockade," *The Verge*, June 18, 2014; Steven Max Patterson, "Mesh Networks and FireChat: How Hong Kong Protestors Are Keeping Communications Alive," *NetworkWorld.com*, October 2, 2014.

by criminal groups (terrorists or organized crime), these groups are more likely to employ more traditional wireless networks because mesh networks are much less practical for operating a fixed, long-term, high-bandwidth communications network. Although communications are sometimes encrypted between relay machines (e.g., using HTTPS), network traffic is generally observable to the relaying host. This may reduce the incentive for one user to connect with untrusted users.

In summary, these technologies partially pass the first test (communication), fail the second (material advantage to criminals), and pass the third test (openly available). Therefore, we conclude that the mesh networking technologies funded under the DRL Portfolio are not more likely to be used for illicit purposes, relative to non-DRL solutions.

Proxy/VPN

Both proxy and VPN technologies enable anti-censorship and anti-surveillance capabilities by acting as intermediaries to a user's Internet communication. Some tools achieve this by relaying (proxying) a censored user's traffic through a single, central server located outside of the censoring regime, while other tools employ more-sophisticated techniques to manage and distribute lists of their proxy or VPN servers.[10] Either way, the user's true location is masked by the proxy or VPN service. While there are some similarities with mesh network technologies, an important difference is that proxy and VPN solutions require an existing Internet connection to function. The tools funded under the DRL Portfolio assist human rights activists by enabling them to access online content (e.g., websites, information) that otherwise would be blocked or surveilled.

We now apply the three-part test for illicit use. First, the proxy and VPN tools funded under the DRL Portfolio would only partially solve the communication, identification, and advertising problems because

---

[10] While both of these are "one-hop" solutions, the main difference between VPNs and proxies is that VPNs often maintain persistent connections between the user and the VPN provider and tunnel all of the computer's communication through an encrypted tunnel. Proxy services, on the other hand, often just redirect a user's web traffic to the proxy server.

they simply enable network connectivity (access), rather than provide a platform to sell illegal goods. For instance, these tools could not, in and of themselves, be used to store or advertise information or files.

Second, while the ability to redirect one's Internet messages through an encrypted channel can afford significant security and privacy to any user, the tools funded under the DRL Portfolio would not provide a material advantage to criminals beyond what is already available from many other proxy and VPN software programs.[11] For example, there are many alternative VPN services not funded by DRL that operate in foreign countries, are not subject to U.S. law, and would therefore likely be more desirable to criminals. Two Russian VPN services, for example, are cryptovpn.com and vpn-service.us.[12]

Third, the tools funded under the DRL Portfolio are freely available online, and many of them are simple to install and configure. However, there are a number of safeguards that may deter criminal use. For example, even though some tools encrypt the communication between the user and VPN provider, the user's communication may still be observable to the operators of the service. Indeed, this is what enables operators to block certain kinds of objectionable traffic and restrict some forms of illicit activity. One operator censors about 3 percent of traffic because of what it considers to be objectionable content.[13] In addition, U.S.-based proxy and VPN providers must comply with law enforcement subpoenas and warrants that could reveal the identity of the user or contents of the user's communication.[14] Further, the IP addresses of the proxy/VPN servers themselves may be publicly known, and therefore susceptible to censorship. In addition, a number of the funded tools in this category are able to avoid censorship by

---

[11] This is not to say that they do not provide a material benefit to human rights activists.

[12] Max Goncharov, *Russian Underground Revisited*, Trend Micro, Cybercriminal Underground Economy Series, 2014.

[13] This estimate is based on the operator's own filters for objectionable (not necessarily illegal) content.

[14] For example, in one case, a UK-based VPN provider complied with a request by the U.S. Federal Bureau of Investigation to provide the identity of an alleged criminal. See "HMA VPN User Arrested After IP Handed Over to the FBI," Hacker10.com, September 28, 2011.

routing messages through known friends and colleagues. Given that few human rights advocates would not knowingly facilitate criminal activity, this design feature suggests that these tools would be of greater practical use to human rights activists, rather than criminals. However, despite these and other safeguards that exist for proxy and VPN software, criminal use cannot be completely dismissed.

In summary, proxy/VPN technologies partially pass the first test (communication), fail the second test (material advantage to criminals), and pass the third test (openly available). Therefore, we conclude that the proxy/VPN tools funded under the DRL Portfolio are not more likely to be used for illicit purposes, relative to alternative (non–DRL-funded and non–U.S.-based) solutions.

Secure Mobile Communication

The secure mobile communication technologies funded under the DRL Portfolio provide encrypted video, voice, and/or text messaging for mobile devices (iOS and Android). They are used either to replace or augment unencrypted services such as Skype, Google Talk, Jabber, and Facebook. Some technologies also provide secure storage for voice and text messages. In addition, others provide secure messaging for PC, Mac, and Linux platforms. Other mobile technologies funded under the DRL Portfolio facilitate delay-tolerant networking, which is necessary in situations where basic Internet connectivity is unreliable and may be unavailable for hours or days. For example, consider a protest where individuals record the surrounding activities, but then must wait for days before communicating the events to others. These mobile applications help ensure that voice, text, email, and video messages are transmitted reliably and securely.

These secure mobile technologies can be a critical enabler of human rights activities. Using the microphones and cameras available on these devices, activists are able to document and disseminate abuses by corrupt officials, military and police forces, drug cartels, or other violent actors. Secure mobile technologies also provide a means to safely store these images and messages when they are captured, transmit the images when bandwidth is available, and even alert allies when something goes wrong, all while protecting the user and the device from reprisal.[15]

We now apply the three-part test for illicit use. First, these projects can solve a criminal's communication problem. Fundamentally, they provide a secure means of exchanging messages between two parties who are known to one another (as opposed to broadcasting messages to unknown groups of people, or marketing one's illicit services).

Second, after evaluating each tool in detail, we conclude that they do not provide a material capability to criminals in excess of what would have existed without DRL funding. There are numerous other technologies, not funded by DRL, that provide similar capabilities for encrypting voice and text messages on mobile devices. For instance, Silent Circle (a for-profit company) sells a specialized phone with a custom-built operating system designed specifically to provide strong encryption for voice, video, text, and web communication.[16] Mobile applications such as Wikr[17] also provide secure messaging, and it is known that, in some cases, organized criminals develop and use their own instant messaging technology.[18]

Third, these DRL-funded tools are free of cost, and many are available as both mobile applications and source code. However, there are a number of safeguards that may restrict some illicit use. Despite the secrecy that is afforded by encrypting text or voice communication, metadata can still reveal the source and destination of messages sent from these devices,[19] thereby facilitating law enforcement surveillance and investigation. Further, microphones and cameras commonly available on these mobile platforms present significant risks to criminals if

---

[15] Tanya O'Carroll, "Mobile Technologies Helping Activists and Human Rights Defenders," *Ethical Consumer*, undated.

[16] Blackphone, homepage, undated.

[17] Wickr, "How Wickr Works," undated.

[18] Jeremy Kirk, "Hackers Build Private IM to Keep Out the Law," *ComputerWorld.com*, March 28, 2007.

[19] Either from the IP address or from the device's unique identification number.

they can be enabled surreptitiously and remotely by law enforcement to record criminal activity.

In summary, these technologies partially pass the first test (communication), fail the second test (no material advantage to criminals), and pass the third test (openly available). Therefore, we conclude that the technologies funded under the DRL Portfolio are not more likely to be used for illicit purposes, relative to non-DRL solutions.

Tor (The Onion Router)

DRL directly and indirectly funds a number of distinct Tor subprojects that together improve Tor's technology, usability, documentation, and customer support.[20] Overall, the Tor application provides both strong anonymity and security for Internet-based communication using a distributed architecture and a custom, open-source protocol to encapsulate and tunnel Internet traffic. A user's network traffic passes through separate relays (which could be located anywhere in the world) within the Tor network, adding a layer of encryption with each connection, such that any one Tor relay is able to observe only the source or destination of traffic for Tor servers with which it directly communicates. For example, in a three-hop circuit, the middle server cannot observe the IP address of the user making the initial request, nor the final destination of the request. This protocol provides a strong defense against traffic analysis, censorship, and privacy intrusions. Conditional on the user not leaking personal information, anonymity is assured not only within a single session (i.e., a website would be unable to identify the source IP address of the user), but also between sessions (i.e., two websites would be unable to associate a user across two sessions).[21] These characteristics make Tor highly desirable for anyone seeking strong confidentiality

---

[20] To be clear, the Tor Project also receives funding from individuals and organizations, including over 4,300 personal donations, as well as other commercial and federal government agencies. See The Tor Project, "Tor: Sponsors," web page, undated b.

[21] Of course, this is barring user-side tracking through cookies or other mechanisms.

through encrypted communication and strong anonymity (unlinkability) between the user and the destination website.

The digital protections afforded by Tor provide many benefits to netizens and human rights activists alike. For example, journalists use Tor to communicate with dissidents and whistleblowers. Individuals from minority and disadvantaged groups use Tor to communicate with others suffering the same threats and victimizations. One commenter wrote,

> Even though I live in a nordic [sic] country I use Tor for writing blogs online. My views are not very popular among my peers and I could easily loose [sic] my job. Freedom of expression is not guaranteed even in a democracy. It's something we have to work for every day. Tor helps to ensure we can speak our minds without fear.[22]

Another commenter wrote,

> As an activist for transgender rights I'm frequently contacted by trans people globally. In particular, for some reason my name has become known among trans people in the middle east and south asia [sic]. I frequently encourage my contacts to use tor [sic] to protect themselves.[23]

Law enforcement and intelligence agencies also use Tor to track and apprehend criminals and interrupt threats to national security.[24] Undercover police officers require the kinds of anonymity and confidentiality that only technologies like Tor can provide. For example, one police officer remarked that he used Tor when working on cases related to Internet crimes against children. He would use Tor to connect to social media sites to help maintain his undercover identity when communicating with those alleged to participate in child exploitation.

---

[22] Anonymous comment on the Tor Project blog (The Tor Project, "We Need Your Good Tor Stories," blog post, August 17, 2011.)

[23] Anonymous comment on the Tor Project blog (The Tor Project, 2011).

[24] See The Tor Project, "Abuse FAQ," web page, undated a.

Similarly, he would use Tor when sending instant messages to alleged narcotics criminals. In addition, law enforcement uses the anonymity provide by Tor to communicate with whistleblowers and other sensitive sources and contacts. Further, the ability to mask the source location of one's computer is critical when investigating the websites of an alleged criminal, whether located within the United States or a foreign country.

Next, we apply the three-part test for illicit use of Tor. In regard to solving a problem as defined by Gambetta,[25] Tor solved a communication problem for the operators and users of the underground contraband marketplace, Silk Road, by allowing them to operate and transact within a webserver that facilitated illegal activities.[26] Further, Tor does not specifically address the challenge of identifying people with common interests, whether criminal or otherwise. However, the anonymity that Tor provides in the underlying communications infrastructure creates a platform where both netizens and criminals feel safer discussing matters that might be more problematic if their identifies were revealed. Regarding the problem of advertisement, Tor is not specifically designed to solve this problem, but the anonymity provided lends itself to advertisement of goods and services that users fear might be prohibited and/or investigated if their identities were to be revealed. (We discuss Tor hidden services later in this chapter.)

Second, while Tor does provide many sophisticated privacy and anonymity features, even relative to non-DRL funded tools, there do exist other technologies that provide similar capabilities as Tor that are not funded by DRL. For example, a number of recent trends suggest that web-based social media platforms (such as Instagram and Kik) are heavily used for activities similar to those conducted on the Tor network, and that they are even more attractive than Tor because of the number of potential buyers in the underground marketplace (e.g., Silk Road). For example, posting photos on Instagram of guns and drugs under usernames such as "ihavedrugs4sale" or using hashtags like

---

[25] Gambetta, 2011.

[26] "The Amazons of the Dark Net," *The Economist*, November 1, 2014.

"#ar15" solves the challenges of advertising goods and being discovered by individuals with similar interests.[27]

In addition, the FreeNet service (which provides distributed and encrypted file storage and retrieval) is designed to protect the identity of the user requesting content and the physical location of that content, and may well be preferred to Tor when solving the problem of criminals seeking to advertise stolen goods.[28] I2P is also similar to Tor, in that it provides encrypted network communication that can be used to access public (and private) Internet services anonymously.[29] That being said, online black-marketplaces that leverage the Tor network are reportedly still prospering.[30]

Nevertheless, Tor *preceded* DRL involvement by over a decade and, therefore, core capabilities would have existed with or without DRL support.[31] Indeed, the fundamental Tor technology was originally sponsored by the U.S. Naval Research Laboratory in the mid-1990s and predates any DRL funding. In addition, The Tor Project is a large software development project consisting of many separate but related components, only some of which are funded by DRL. Indeed, the specific components that are funded by DRL generally relate to internationalization and usability of Tor applications, rather than core components that would assist criminal activity.

Third, by design, Tor software and its services are freely available online. However, there are a number of safeguards that could restrict illicit use. For example, criminals face some of the same challenges as proxy/VPN users—that the IP addresses of Tor relay nodes can become known, which would enable governments, law enforcement,

---

[27] Fletcher Babb, "Lean on Me: Emoji Death Threats and Instagram's Codeine Kingpin," *Vice.com*, October 24, 2013.

[28] See https://freenetproject.org/.

[29] See https://geti2p.net/en/.

[30] "The Amazons of the Dark Net," 2014.

[31] According to the Tor project, DRL funding began in 2013, while Tor development began in the mid-1990s and Tor was first released in 2002. For more information on Tor funding, see The Tor Project, undated b. For more information on initial release, see Roger Dingledine, "Pre-Alpha: Run an Onion Proxy Now!" email dated September 20, 2002.

and even for-profit websites (e.g., Google) to track and block requests.[32] Further, the actions of users conducting illicit activity over Tor can still be revealed through human failure and conventional law enforcement tactics. For instance, despite using Tor to email a bomb threat to Harvard University, a student was caught because police were able to quickly narrow the list of those people on the Harvard University network who were using Tor at the time the email was sent.[33] Despite operating within the Tor network, the Federal Bureau of Investigation's (FBI's) capture of the online criminal marketplace, Silk Road (as well as its successor, Silk Road v2.0), was also accomplished in part by conventional law enforcement techniques.[34] Finally, there is also evidence that Tor systems, like all software applications, suffer from software attacks and abuse. One news story reported that a Tor exit node had been maliciously modifying downloads of the actual Tor software to inject malware and possibly identify the activity (illicit or otherwise) of its users.[35] There are also indications by researchers that Tor users could be uniquely identified, given sufficient resources available to an adversary.[36]

The Tor Project's leaders also engage in many safeguard activities designed to help promote the appropriate and legitimate use of their technology. For example, much effort is devoted to addressing usability and localization features that help ensure that human rights users in

---

[32] The relatively new feature of Tor Bridges helps mitigate this risk, however.

[33] Runa A. Sandvik, "Harvard Student Receives F for Tor Failure While Sending 'Anonymous' Bomb Threat," *Forbes*, December 18, 2013.

[34] Kim Zetter, "How the Feds Took Down the Silk Road Drug Wonderland," *Wired.com*, November 18, 2013. Regarding Silk Road v2.0, see Federal Bureau of Investigation, New York Field Office, "Operator of Silk Road 2.0 Website Charged in Manhattan Federal Court," press release, November 6, 2014.

[35] Darren Pauli, "Tor Exit Node Mashes Malware into Downloads," *The Register*, October 27, 2014.

[36] Sambuddho Chakravarty, Marco V. Barbera, Georgios Portokalidis, Michalis Polychronakis, and Angelos D. Keromytis, "On the Effectiveness of Traffic Analysis Against Anonymity Networks Using Flow Records," in Michalis Faloutsos and Aleksander Kuzmanovic, eds., *Passive and Active Measurement: Proceedings of 15th International Conference, PAM 2014*, Los Angeles, Calif.: Springer, March 10–11, 2014.

more remote areas have access to privacy tools. In addition, they have invested a great deal of effort in making usage and performance data publicly available for the purpose of research and transparency. For example, it was these data that provided evidence showing surges in user activity during revolutions in the Middle East. It was also these data that allowed operators to observe sudden changes in usage, which was likely due to malicious (botnet) activity.[37] Such data and privacy-preserving efforts help monitor and counter malicious traffic. Finally, the Tor Project's leaders are active in the privacy and security community, including Internet freedom groups, not only within the United States but across the world. They are clear in promoting Tor as a privacy and security tool that can be used by journalists, netizens, law enforcement, U.S. armed forces, and others.[38] For example, Tor representatives have participated in FBI and international law enforcement conferences to educate officials on Tor's capabilities as well as the benefits for their officers when conducting investigations.[39] They are also regularly invited by, and speak with, domestic and international law enforcement agencies to educate and train users regarding the benefits and advantages of secure and anonymous communication.

Next, in the particular case of Tor, we examine realworld usage data as collected by various sources. Although it would be troubling if most Tor traffic were used to conduct illicit activity, this is not supported by available data. For example, Chaabane and colleagues (2010) examined 373 gigabytes of Tor traffic from six Tor exit nodes, distributed globally, for a period of 23 days from late 2009 to early 2010.[40] They found that about 52 percent of the traffic volume was P2P file-sharing (e.g., BitTorrent), with one-half sent unencrypted end-to-end.

---

[37] The Tor Project, "How to Handle Millions of New Tor Clients," blog post, September 5, 2013b.

[38] For full descriptions of Tor users, see The Tor Project, "Inception," web page, undated c.

[39] The Tor Project, "Trip Report, Tor Trainings for the Dutch and Belgian Police," blog post, February 5, 2013a; The Tor Project, "Trip Report, October FBI Conference," blog post, December 16, 2012.

[40] Abdelberi Chaabane, Pere Manils, and Mohamed Ali Kaafar, "Digging into Anonymous Traffic: A Deep Analysis of the Tor Anonymizing Network," in *Proceedings of the 2010 Fourth International Conference on Network and System Security*, September 2010.

A further 36 percent was found to be unencrypted web traffic, 5 percent was encrypted web traffic, 0.25 percent was instant messaging, and the rest was miscellaneous and cleartext protocols. Quite likely, the unencrypted web traffic was not used for illicit purposes; it is unclear what percentage of the instant messaging traffic was illicit—regardless, the overall percentage would be very small. These findings of high proportions of cleartext web traffic (58 percent) and BitTorrent traffic (40 percent) from 709 gigabytes of Tor traffic are supported by similar research.[41] By comparison, estimates from 2011 suggest that web traffic makes up 38 percent of all U.S. Internet bandwidth, while P2P file-sharing composes 19 percent of all bandwidth.[42] Summaries of these results are shown in Figure 7.1.

If a substantial portion of Tor traffic is indeed used for P2P file-sharing, it helps to examine the types of files shared. From a sample of

Figure 7.1
Comparison of Tor and Internet Traffic (by volume)

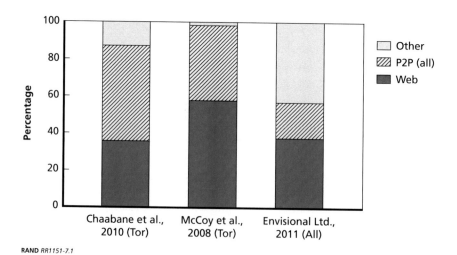

---

[41] Damon McCoy, Kevin Bauer, Dirk Grunwald, Tadayoshi Kohno, and Douglas Sicker, "Shining Light in Dark Places: Understanding the Tor Network," in *Proceedings of the 8th International Symposium on Privacy Enhancing Technologies*, Leuven, Belgium, July 2008.

[42] Envisional, Ltd., *Technical Report: An Estimate of Infringing Use of the Internet*, Cambridge, UK, January 2011, p. 49.

100,000 files on BitTorrent, one report estimates that 85 percent are video files, which consist of pornography (36 percent), films (35 percent), and television programming (13 percent). Software, music, and computer games together comprised another 14 percent of the total BitTorrent traffic. Further, the report identified that about 64 percent of all files were copyrighted material, implying that almost all non-pornographic content was copyrighted. The report authors were not able to determine the percentage of pornographic content that was either illicit or copyrighted.[43]

When examining the types of websites visited by Tor users, Chaabane and colleagues (2010) found results that are reproduced in Table 7.1.

These results suggest that of all the observed Tor activity, only 11.5 percent of destination websites were considered pornographic (for

Table 7.1
Most-Visited Website Types of Tor Users

Rank	Category	Percentage
1	Search engines/portals	14.45
2	Pornography	11.50
3	Computers/internet	11.45
4	Social networking	9.52
11	Blogs/web comm.	2.26
13	Streaming media/mp3	1.82
14	Software downloads	1.66
36	Hacking	0.30
40	Political	0.18
42	Illegal/questionable	0.15
52	Illegal/drugs	0.06

SOURCE: Chaabane et al., 2010.

---

[43] Envisional, Ltd., 2011, p. 10.

which only some would be considered illegal) and only 0.21 percent were identified as specifically illegal (according to U.S. law).

Next, we examine the global distribution of Tor users. If policymakers are ultimately concerned about illicit activity conducted by *American* citizens, then it would be of greatest concern if the vast majority of Tor users originated from the United States. Table 7.2 illustrates the top five client distributions from three separate research papers.

Table 7.2 shows the counts and percentages of Tor clients (as measured by unique IP address) connecting to the researchers' Tor entry nodes, by country. The left panel represents 7,571 client requests collected in 2007–2008, the middle panel shows 7,575 requests collected in 2009–2010, and the right panel shows 5,932 requests collected in late 2010. These data suggest that, consistently, most Tor requests originate from Germany, followed by the United States, China, and Italy; U.S. traffic accounted for, at most, 13 percent of all Tor users. This result is also confirmed by data from the Tor Project itself.[44]

In other research, Huber and colleagues (2010) examined HTTP requests only and found that the majority of web requests belonged to social networking, search engine, and file-sharing websites.[45]

Examination of these data (Tor traffic by service, and client country origin) suggests that while the majority of Tor traffic is unencrypted web and P2P file-sharing, only a small fraction (about 13 percent) of all Tor users originate from the United States.

We evaluated the specific features of Tor that were identified as being funded by DRL. Therefore, we necessarily did not specifically address the functionality of Tor hidden services, an important and controversial component of the overall Tor application. Tor's hidden services provides a mechanism for individuals to host Internet applications (such as websites) anonymously within the Tor network—largely hidden from the public and law enforcement.[46] It has been widely

---

[44] The Tor Project, "Top-10 Countries by Directly Connecting Users," database, undated d.

[45] Markus Huber, Martin Mulazzani, and Edgar Weippl, "Tor HTTP Usage and Information Leakage," in *Proceedings of the 11th IFIP TC 6/TC 11 International Conference on Communications and Multimedia Security*, Linz, Austria, May 2010.

[46] I.e., the so-called "dark web."

Table 7.2
Distribution of Tor Client Locations

	Chaabane et al. (2007–2008)			McCoy et al. (2009–2010)			Li et al. (2010)		
	Country	Count	%	Country	Count	%	Country	Count	%
	Germany	2,304	30	Germany	1,114	15	Germany	1,076	18
	China	988	13	USA	970	13	USA	734	12
	USA	864	11	Poland	839	11	Italy	657	11
	Italy	254	3	Romania	583	8	China	469	8
	Turkey	221	3	Russia	553	7	France	356	6
	Other	2,940	61	Other	3,516	46	Other	2,640	45
	Total	7,571	100	Total	7,575	100	Total	5,932	100

SOURCES: Chaabane et al., 2010, p. 19; McCoy et al., p. 72; and Bingdong Li, Esra Erdin, Mehmet Hadi Güneş, George Bebis, and Todd Shipley, "An Analysis of Anonymity Technology Usage," in *Proceedings of the Third International Conference on Traffic Monitoring and Analysis*, Vienna, Austria, April 2011.
NOTE: Emphasis added.

reported that hidden services are used for criminal activity, such as contraband e-commerce websites (e.g., Silk Road) and child exploitation material.[47] Combined with anonymous currencies used for payment, this combination of technologies has proven difficult for law enforcement to contain and detect. Despite Tor hidden services being beyond the scope of this analysis, it is an important capability that is demanding more recent attention because it presents such a difficult challenge to law enforcement, and is therefore worth acknowledging.

In summary, despite the numerous safeguards relating to both the Tor application and its practical use, we conclude that Tor passes the first part of the three-part test (communication, advertising). However, it fails the second test because it does not provide a material capability that would not otherwise be available without DRL funding. It passes the third test because, with only limited safeguards, it is still freely available. Nevertheless, despite the popularity of Tor and its impressive capabilities, we conclude that DRL funding has not made it more likely for this tool to be used for illicit purposes.

Two Other Projects

One project provides a web service and desktop and mobile applications designed to help human rights activists, journalists, and concerned citizens securely collect, document, and share human rights abuses.

Applying the three-part test for illicit use, we observe that because this project is primarily an application that enables secure storage, it could partially address a criminal's communication problem, to the extent that multiple criminals would create and exchange content among a shared account using this application. That is, it would assist in exchanging content between known users.

Second, the capability to store files securely (i.e., encrypted) is a feature easily available to criminals with many other software applications. Indeed, there are alternative, non–DRL-funded applications that

---

[47] Alex Biryukov, Ivan Pustogarov, Fabrice Thill, Ralf-Philipp Weinmann, "Content and Popularity Analysis of Tor Hidden Services," 2013.

provide similar secure file storage and would likely be more convenient for illicit use. These tools are popularly known as cyberlockers, and include MegaUpload, RapidShare, and HotFile.

Third, while this application is freely available, there are a number of safeguards that may limit illicit use of this application. All files stored in the client applications or uploaded to web services are encrypted, preventing the operators of the service from accessing the unencrypted files. However, operators could observe metadata related to account usage and access (e.g., presence of files, IP addresses of account logins), which may thereby become available for law enforcement. This could dissuade criminal use of this tool because it may not provide strong anonymity for criminals looking to conceal their activities or location.

Overall, this project partially passes the first test (communication), fails the second test (material advantage to criminals), and passes the third test (openly available). Therefore, we conclude that this DRL-funded project would not be more likely to be used for illicit purposes, relative to non-DRL solutions.

The second project provides an operating environment that is bootable from a USB drive, and is preconfigured with privacy-preserving software to enable users to browse the Internet securely and anonymously. Surveillance-free Internet connectivity is facilitated by using a secure operating platform that also helps prevent malware from compromising the identity of the user. This tool can also be configured with an encrypted file system or as a read-only file system, thereby preventing the accidental storage of potentially privacy-revealing information.

As we apply the three-part test for illicit use, we observe that because this project is designed to provide a platform for secure and anonymous communications, it does not directly or specifically solve any of the communication, identification, or advertising problems.

Second, while the project as a whole provides features that could facilitate illicit use, the individual components are already available to criminals from non–DRL-funded software. For example, TAILS is an operating environment that provides a bootable operating system preconfigured for anonymous Internet communication.[48] In addition, many Linux operating systems provide live distributions that are bootable from a CD/DVD and therefore protect against surveillance or system compromise from malware.

Third, the distribution of this tool is limited only to individuals who have undergone in-person training with, and have been vetted by, the developers of this project. This clearly provides a strong safeguard against illicit use.

Therefore, because this project fails each of the three tests, we conclude that this DRL-funded project would not be more likely to be used for illicit purposes, relative to non-DRL solutions.

Summary

The conclusions from this analysis are summarized in Table 7.3. Each DRL-funded project is listed by row, with columns 1–3 providing the results of the three-part test. Column 4 summarizes the conclusion by addressing whether DRL funding has increased the likelihood for a tool to be used for illicit purposes. As stated, in order for DRL involvement to produce an affirmative answer, the tool, project, or service must pass each of the three tests. Based on our analysis, we concluded that DRL's involvement in funding these programs has *not* increased their potential to be used for illicit activities. While column 2 shows that none of the tools provides a material advantage to criminals over alternative technologies, this result occurs because of different reasons, which are explained above and summarized in column 5.[49] Specifically, we found that because the digital safety and anti-DDoS programs are provided in person or to customers who support human rights activities and are vetted by the operators, they would not increase the likeli-

---

[48] For more information, see https://tails.boum.org/.

[49] As stated, the rules that compose the three-part test were created to produce the minimum set of conditions that could lead to increased illicit use. While all responses to the second test (Does the tool, project, or service provide a material capability not available absent DRL funding?) were negative, this does not diminish the value of the question because these answers, certainly, were not known ex ante.

Table 7.3
Summary of Evaluations

	Solves a criminal's communication problem? (1)	Provides a material advantage to criminals? (2)	Is the tool reasonably accessible to criminals? (3)	Has DRL's involvement increased the potential for illicit use? (4)	Explanation of result (5)
Digital safety	No	No	No	No	Training is provided in person and does not provide a material capability to criminals
Anti-DDoS	Yes	No	No	No	Customers are vetted by the providers and must support human rights activities
Mesh networks	Yes	No	Yes	No	Many alternative non–DRL-funded tools exist that may better meet the needs of illicit users
Proxy/VPN	Yes	No	Yes	No	Many alternative non–DRL-funded tools exist that may better meet the needs of illicit users
Secure communication	Yes	No	Yes	No	Many alternative non–DRL-funded tools provide similar capabilities that may better meet the needs of illicit users
Tor	Yes	No	Yes	No	Sophisticated capabilities and a large user base; however, Tor pre-dated DRL involvement
Other/Project A	No	No	Yes	No	Alternative non–DRL-funded tools provide similar capabilities that may better meet the needs of illicit users
Other/Project B	No	No	No	No	Users are vetted by the owners of the project

NOTE: This table provides summary results of the analysis described in this report. Specifically, it summarizes the extent to which the programs funded by DRL may or may not have increased the potential for illicit use by criminals.

hood for illicit use. Regarding the mesh network technologies, proxy/VPN tools, and the secure communication tools, there exist a number of alternative tools not funded by DRL that provide similar capabilities and may well be more suitable for criminal activity because, for instance, services that operate in foreign countries are much less likely to comply with U.S. law enforcement requests. Next, even though Tor provides strong encryption and anonymity, these core features predated any funding by DRL. Finally, we find that neither of the additional projects evaluated would increase criminal activity either because there are many alternative tools available to criminals or because potential users are vetted by the operators of the program.

In summary, based on our underlying assumptions and methodology, we conclude that DRL's involvement in funding these privacy and security tools has not increased their likelihood for illicit use.

Note that while we used a methodology and framework that was created to evaluate these specific projects, we believe it is generalizable, such that it can be used to evaluate future projects.

CHAPTER EIGHT

Additional Mitigating Safeguards

Given the variation in the use of the privacy and security technologies and services evaluated within this report, the challenges faced by governments seeking to fund Internet freedom tools are twofold. First, governments could argue that the benefits of enhancing freedom (e.g., for netizens) outweigh the costs of enabling freedom for criminals to escape prosecution. Second, they could argue that their projects can be, and are, designed to tilt the benefits toward increasing human rights and against illicit activity. In effect, the first challenge tests the benefits of supporting these tools against their costs, while the second suggests that there are ways to refine these programs to minimize unintended consequences.

The second challenge, however, rests on three related propositions: First, the ways these tools support the exercise of human rights are different from the ways these tools support illicit activity; second, the criteria that predispose human rights activists to use DRL tools differ from the criteria that predispose criminals to use similar tools already on the market; and third, safeguards can be established that would deter illicit use. It is this third matter—the potential for additional safeguards to restrict or deter criminal use—that we discuss further in this chapter.

In evaluating the projects funded under the DRL Portfolio, we have identified many safeguards and designs that could limit and restrict their use by criminals. In some cases, the projects were available only to a limited group of known individuals who are vetted by the project owners and operators. In other cases, the capability for lawful investigation provides a strong deterrent against criminal use. Further, it was shown in the case of Tor that even the most sophisticated tools are still susceptible to human error and law enforcement investigation.

One way to assuage concerns regarding illicit use of DRL-funded tools is to encourage broader localization, digital safety training, and awareness efforts by more grantees in more countries around the world. DRL could also request that each of its grantees document the safeguards, designs, assumptions, and other factors that would limit, restrict, or deter use of its technologies by criminals. In addition, it could request that each grantee observe and document (in an appropriate manner) any evidence of illicit use of its tool.

Clarity in the debate over circumvention tools may come through greater transparency by law enforcement (and perhaps the intelligence community) in reporting the number of times that its efforts are thwarted by privacy-enhancing technologies during investigations. Such public disclosure already characterizes law enforcement's use of wiretaps, at least with regard to the Wiretap Act, which requires the government to publish statistics regarding their use.[1] For instance, according to the U.S. Administrative Office of the Courts, 41 of the 3,576 wiretaps issued in 2013 employed encryption, and in only nine cases did the encryption prevent law enforcement from deciphering the messages.[2] Further, 87 percent of wiretaps involved drug offenses, and 97 percent of all wiretaps from 2013 involved portable devices. However, this provides only a limited view of the difficulty that law enforcement faces, since the Wiretap Act governs only information collected at the time of transmission. The Pen Register Act,[3] on the other hand, governs metadata, which may be more useful as more messages are sent by mobile and electronic devices. However, the Pen Register Act

---

[1]   United States Code, Title 18, Chapter 119, Wire and Electronic Communications Interception and Interception of Oral Communications (§§ 2510–2522).

[2]   These relate to oral, wire, and electronic communications. See United States Courts, *Wiretap Report 2013*, December 31, 2013. However, note that it is likely that these data underestimate the true use of encryption by an unknown amount.

[3]   United States Code Title 18, Chapter 206, Pen Registers and Trap and Trace Devices (§§ 3121–3127).

(even as modified by the Patriot Act of 2001) does not require public disclosure of any collection statistics. Therefore, it is possible that additional data could help determine the extent to which anti-censorship and anti-surveillance tools facilitate criminal activity.

Further, while we already identified some variation between netizen and criminal preferences for these projects, additional research could be conducted to more fully understand these differential preferences (i.e., circumvention, encryption, access, usability, education). This could enable DRL to tailor its awards to those grantees that develop capabilities of more interest to netizens and of less potential interest to criminals. In addition, research devoted to better understanding these preferences (and constraints) could also uncover a better combination and application of safeguards to be implemented across a portfolio of grantees.

Finally, very little is currently known about the extent to which various kinds of traffic (legal and illicit) pass through privacy and security tools in general, and specifically the tools evaluated in DRL's Portfolio. Research that examines and quantifies these network behaviors in a legal[4] and privacy-preserving manner could greatly inform this critical policy debate concerning the illicit use of Internet freedom tools.[5]

---

[4] For example, while large-scale analysis of VPN and Tor traffic is empirically possible, care must be taken not to violate U.S. wiretap laws, such as the Electronic Communications Privacy Act (18 U.S.C. §§2511–2522).

[5] For one commenter's discussion of ethical research in this area, see Christopher Soghoian, "Enforced Community Standards for Research on Users of the Tor Anonymity Network," *Financial Cryptography and Data Security, Lecture Notes in Computer Science*, Vol. 7126, 2012.

CHAPTER NINE

Conclusion

This report examined the extent to which the Internet freedom projects funded by the U.S. Department of State could be used for illicit purposes. It first examined the benefits of these projects in promoting DRL's mission to promote Internet freedom across the world for human rights purposes. Then, it examined the projects' potential to be used for illicit purposes by applying a three-part test: Do they solve a criminal's communication problem? Do they provide a material advantage to criminals? Are they reasonably accessible to criminals?

Although Congress has expressed its concern that Internet freedom technologies not be used for illicit purposes, the program has been robustly funded for the past several years, indicating a bipartisan consensus that it can be a useful element in U.S. foreign policy. For instance, congressional funding for DRL has increased over the past few fiscal years, despite budgetary constraints imposed by the Budget Control Act of 2011 and deep cuts to several foreign assistance programs. In fact, Congressional appropriations have surpassed the amount requested in the president's budget in recent fiscal years, which indicates strong support for DRL's mission, with operational appropriations increasing from $18.8 million in fiscal year 2009 to $32.3 million in fiscal year 2015.[1] This support is also reflected in corresponding increases in the number of DRL full-time employees, which has

---

[1] U.S. Department of State, *Congressional Budget Justification for Fiscal Year 2011, Volume 1: Department of State Operations*, February 1, 2010, p. 351; "Explanatory Statement Submitted by Mr. Rogers of Kentucky, Chairman of the House Committee on Appropriations, Regarding the House Amendment to the Senate Amendment on H.R. 83, Consolidated

increased from 118 in fiscal year 2009 to 161 in fiscal year 2015.[2] The Human Rights & Democracy Fund, referred to by the State Department as the "Bureau of Democracy, Human Rights and Labor's flagship program,"[3] has seen steady increases in levels of funding over the past several fiscal years.

In addition to the analysis provided earlier in this report, we offer additional insights into the complex issue of illicit use of privacy and security technologies. First, as has often been stated, technology is agnostic to its use. Whether one considers a car, a pen, the telephone, or software that enables access to the public Internet, there is little opportunity to enable legitimate use while simultaneously preventing illicit use. When addressing this potential for dual use, as one software developer appropriately stated, "that was never its intended purpose, but there is no way to prevent any specific usage of such tools without sacrificing all its advantages."[4] This developer was referring to the problem that, in order to filter illicit (or any sort of objectionable) content from a communication software program, it must be capable of inspecting and vetting each message. This inspection capability, while potentially noble, in effect introduces the very privacy invasion that the software may have been designed to prevent.

It is also useful to consider the possibility that criminal groups employ many different technologies, not funded through DRL's Portfolio, to communicate online and, very often, these are the same technologies that everyday Americans use. For example, a recent article stated that "there are plenty of mainstream technologies that criminals can use to hide their activities: satellite phones, PIN [peronal identifi-

---

and Further Continuing Appropriations Act, 2015," *Congressional Record*, Vol. 160, No. 151, December 11, 2014, p. H9948.

[2]   U.S. Department of State, *Congressional Budget Justification for Fiscal Year 2010*, May 12, 2009, p. 367; U.S. Department of State, *Congressional Budget Justification for Fiscal Year 2016, Appendix 1: Department of State Diplomatic Engagement*, February 2, 2015, p. 237.

[3]   U.S. Department of State, Bureau of Democracy, Human Rights, and Labor, "DRL Programs," web page, undated.

[4]   "Interview with Bernd Kreuss of TorChat," Free Software Foundation, August 26, 2013.

cation number] messaging on BlackBerrys and even Apple iMessage."[5] Conversations with security and Internet freedom experts concluded that criminals also use Skype, instant messaging, Internet relay chat,[6] Reddit, Facebook, encrypted email, and stolen cell phones. A recent report further described black-market techniques that include everything from bulletin board messaging to email exchange (including draft email dead drops), web forums, and private Twitter accounts.[7] According to another report, criminals also use Instagram (an online photo- and video-sharing service) to buy and sell drugs online,[8] and the anonymous question-answer website ask.fm is being used to solicit and respond to daily concerns of would-be religious extremists.[9]

There is also evidence that foreign countries and religious extremist groups (e.g., jihadists) are distrustful of any software written or supported by Western (especially American) developers.[10] This would be true regardless of whether the software is open- or closed-source, or whether the software better helps protect their identity and conceal their Internet traffic. These suspicions would greatly reduce the incentive for terrorists (or other foreign actors) to use any of the tools funded by the DRL Portfolio. Further evidence of this is provided by news articles cataloging the number and variety of encryption tools that have been custom-developed by extremist groups to provide anony-

---

[5] Lev Grossman and Jay Newton-Small, "The Secret Web: Where Drugs, Porn, and Murder Live Online," *Time*, November 11, 2013.

[6] Norton by Symantec, "The Cybercrime Blackmarket," web page, undated.

[7] Lillian Ablon, Martin C. Libicki, and Andrea A. Golay, *Markets for Cybercrime Tools and Stolen Data: Hackers' Bazaar*, Santa Monica, Calif.: RAND Corporation, RR-610-JNI, 2014.

[8] Fletcher Babb, "How Instagram's Drug Deals Go Undetected," *VentureBeat.com*, September 11, 2014.

[9] John Hall, "'U dnt need much, u get wages here, u get food provided and place to stay': The Rough Travel Guide British ISIS Fighters Are Using to Lure Fellow Britons in to Waging Jihad in Iraq," *Daily Mail*, June 18, 2014.

[10] Rodrigo Bijou, "An Overview of Jihadist Encryption Programs," blog post, October 31, 2013.

mous text and messaging capabilities on mobile and PC platforms.[11] For example, Al-Fajr and the Global Islamic Media Front (two arms of al Qaeda) are each reportedly developing separate encryption software for the Android platform.[12] Organized criminals were also reportedly developing their own instant messaging software to exchange stolen financial information.[13]

To further determine whether the DRL Portfolio is fostering criminality, consider two worlds: today's DRL-funded world, and a hypothetical world in which the DRL Portfolio never existed. Many of these projects exist in both worlds. However, improvements that have been funded through the DRL program exist only in the former world. For example, the Tor Project began in the mid 1990s within the U.S. Naval Research Laboratory, but is now funded by many private and public entities. In 2012, only 28 percent of its federal funding came from the Department of State.[14] While it is legitimate to inquire whether these improvements have made it more likely for the tool to be used for criminal purposes, we posit that it is not legitimate, by virtue of this analysis, to hold DRL accountable for all criminal use when these tools would have existed absent DRL's funding.

Further, when choosing which efforts to fund, DRL is highly cautious and clear with its effort to support training and technologies that support human rights and Internet freedom around the world:

> Significant efforts are made to avoid supporting those who advocate for violence or other activities that violate or impair the enjoyment of others' human rights. . . . Internet freedom technologies funded by State and USAID [U.S. Agency for International Development] are designed for deployment in repressive environ-

---

[11] Recorded Future, 2014a, and Bijou, 2013, respectively.

[12] David Kravets, "Terrorists Embracing New Android Crypto in Wake of Snowden Revelations," *Ars Technica*, August 1, 2014.

[13] Kirk, 2007.

[14] Moody, Famiglietti, and Andronico, LLP, *The Tor Project, Inc. and Affiliate: Consolidated Financial Statements and Reports Required for Audits in Accordance with Government Auditing Standards and OMB Circular A-133—December 31, 2013*, Tewksbury, Mass., July 11, 2014, p. 12.

ments, and designed based on the feedback and input of activists, bloggers, and others who are working in those environments in order to meet their needs. Distribution methods and networks for internet freedom technologies reflect this, and focus on helping individuals in Internet repressive environments. In sensitive countries, program participants are vetted to avoid funding terrorists or intelligence operatives.[15]

DRL also states within its annual solicitation that it "supports programs that uphold democratic principles, support and strengthen democratic institutions, promote human rights, and build civil society around the world," and that

> DRL will not consider projects that reflect any type of support for any member, affiliate, or representative of a designated terrorist organization, whether or not elected members of government. Organizations that are invited to submit proposals and subsequently approved for an award may be required to submit additional information on the organization and key individuals for vetting.[16]

Finally, as previously described in this report, while the projects supported under the DRL Portfolio may not provide a material capability for criminals, this does not suggest that they do not provide a critical service for their intended audience—human rights activists and other at-risk groups across the world. Specifically, it is likely true that reducing or eliminating funding for these projects would disproportionally harm human rights activists for the simple reason that criminals have a greater selection of security and privacy tools available to them to conduct illicit activity. They can use any of the tools described here or any free or commercial product, as well as any custom-built or illegal tools. However, human rights activists, who are less well-funded, are much more limited in their selection of technologies that

---

[15] Email to authors from DRL staff, November 18, 2013.

[16] U.S. Department of State, "DRL Internet Freedom Annual Program Statement for Internet Freedom Technology," web page, April 3, 2013.

preserve their privacy and safety. Removing these legitimate technologies could leave them without a secure means of communication and impose upon them a serious risk of personal harm and persecution.[17]

---

[17] This argument was first suggested to the authors by members of the Tor community.

References

Ablon, Lillian, Martin C. Libicki, and Andrea A. Golay, *Markets for Cybercrime Tools and Stolen Data: Hackers' Bazaar*, Santa Monica, Calif.: RAND Corporation, RR-610-JNI, 2014. As of June 9, 2015:
http://www.rand.org/pubs/research_reports/RR610.html

"The Amazons of the Dark Net," *The Economist*, November 1, 2014. As of June 9, 2015:
http://www.economist.com/news/international/21629417-business-thriving-anonymous-internet-despite-efforts-law-enforcers

Babb, Fletcher, "Lean on Me: Emoji Death Threats and Instagram's Codeine Kingpin," *Vice.com*, October 24, 2013. As of June 9, 2015:
http://noisey.vice.com/blog/lean-on-me

———, "How Instagram's Drug Deals Go Undetected," *VentureBeat.com*, September 11, 2014. As of June 9, 2015:
http://venturebeat.com/2014/09/11/how-instagrams-drug-deals-go-undetected/

Bijou, Rodrigo, "An Overview of Jihadist Encryption Programs," blog post, October 31, 2013. As of June 9, 2015:
http://www.rbijou.com/2013/03/18/an-overview-of-jihadist-encryption-programs/

Biryukov, Alex, Ivan Pustogarov, Fabrice Thill, Ralf-Philipp Weinmann, "Content and Popularity Analysis of Tor Hidden Services," 2013. As of June 16, 2015:
http://arxiv.org/abs/1308.6768

Blackphone, homepage, undated. As of June 30, 2014:
https://www.blackphone.ch/

Brandom, Russell, "Iraqis Seek Out New Tools to Blast Through Internet Blockade," *The Verge*, June 18, 2014. As of June 30, 2014:
http://www.theverge.com/2014/6/18/5820694/iraqis-seek-out-new-tools-to-blast-through-internet-blockade

Chaabane, Abdelberi, Pere Manils, and Mohamed Ali Kaafar, "Digging into Anonymous Traffic: A Deep Analysis of the Tor Anonymizing Network," in *Proceedings of the 2010 Fourth International Conference on Network and System Security*, September 2010, pp. 167–174.

Chakravarty, Sambuddho, Marco V. Barbera, Georgios Portokalidis, Michalis Polychronakis, and Angelos D. Keromytis, "On the Effectiveness of Traffic Analysis Against Anonymity Networks Using Flow Records," in Michalis Faloutsos and Aleksander Kuzmanovic, eds., *Passive and Active Measurement: Proceedings of 15th International Conference, PAM 2014*, Los Angeles, Calif.: Springer, March 10–11, 2014, pp. 247–257.

De Filippi, Primavera, "It's Time to Take Mesh Networks Seriously (and Not Just for the Reasons You Think)," *Wired.com*, January 2, 2014. As of July 14, 2014:
http://www.wired.com/2014/01/
its-time-to-take-mesh-networks-seriously-and-not-just-for-the-reasons-you-think/

Dingledine, Roger, "Pre-Alpha: Run an Onion Proxy Now!" email dated September 20, 2002. As of June 12, 2015:
http://archives.seul.org/or/dev/Sep-2002/msg00019.html

Edge Velocity Corporation, "About Us," web page, undated. As of June 30, 2014:
http://www.edgevelocity.com/about_us.html

Envisional, Ltd., *Technical Report: An Estimate of Infringing Use of the Internet*, Cambridge, UK, January 2011. As of June 9, 2015:
http://documents.envisional.com/docs/Envisional-Internet_Usage-Jan2011.pdf

"Explanatory Statement Submitted by Mr. Rogers of Kentucky, Chairman of the House Committee on Appropriations, Regarding the House Amendment to the Senate Amendment on H.R. 83, Consolidated and Further Continuing Appropriations Act, 2015," *Congressional Record*, Vol. 160, No. 151, December 11, 2014.

Faris, Robert, John Palfrey, Ethan Zuckerman, Hal Roberts, and Jillian York, *International Bloggers and Internet Control: Full Survey Results*, Cambridge, Mass.: Harvard University Berkman Center for Internet and Society, August 18, 2011. As of January 23, 2015:
https://cyber.law.harvard.edu/publications/2011/
International_Bloggers_Internet_Control_Full_Survey_Results

Federal Bureau of Investigation, New York Field Office, "Operator of Silk Road 2.0 Website Charged in Manhattan Federal Court," press release, November 6, 2014. As of June 9, 2015:
http://www.fbi.gov/newyork/press-releases/2014/
operator-of-silk-road-2.0-website-charged-in-manhattan-federal-court

Forde, Patrick, and Andrew Patterson, "Paedophile Internet Activity," *Australian Institute of Criminology: Trends and Issues in Crime and Criminal Justice*, Vol. 97, November 1998. As of January 25, 2015:
http://pandora.nla.gov.au/pan/10850/20110125-1520/
%7B8DC57715-E250-43B1-91BD-D04752499CA8%7Dti97.pdf

Gambetta, Diego, *Codes of the Underworld: How Criminals Communicate*, Princeton, N.J.: Princeton University Press, 2011.

Goncharov, Max, *Russian Underground Revisited*, Trend Micro, Cybercriminal Underground Economy Series, 2014. As of June 30, 2014:
http://www.trendmicro.com/cloud-content/us/pdfs/security-intelligence/white-papers/wp-russian-underground-revisited.pdf

Gorman, Siobhan, "Iran-Based Cyberspies Targeting U.S. Officials, Report Alleges," *Wall Street Journal*, May 29, 2014. As of July 2, 2014:
http://online.wsj.com/articles/
iran-based-cyberspies-targeting-u-s-officials-report-alleges-1401335072

Grossman, Lev, and Jay Newton-Small, "The Secret Web: Where Drugs, Porn, and Murder Live Online," *Time*, November 11, 2013.

Hall, John, "'U dnt need much, u get wages here, u get food provided and place to stay': The Rough Travel Guide British ISIS Fighters Are Using to Lure Fellow Britons in to Waging Jihad in Iraq," *Daily Mail*, June 18, 2014. As of June 9, 2015:
http://www.dailymail.co.uk/news/article-2661177/Travel-light-leave-Islamic-books-home-dont-arouse-suspicion-Isis-militants-offer-travel-advice-jihadists-arriving-Syria-Iraq-Britain.html

Harris, Shane, "Exclusive: Inside the FBI's Fight Against Chinese Cyber-Espionage," *Foreign Policy*, May 27, 2014. As of July 2, 2014:
http://www.foreignpolicy.com/articles/2014/05/27/
exclusive_inside_the_fbi_s_fight_against_chinese_cyber_espionage

Henry, Ryan, Stacie L. Pettyjohn, and Erin York, *Portfolio Assessment of Department of State Internet Freedom Program: An Annotated Briefing*, Santa Monica, Calif.: RAND Corporation, WR-1035-DOS, 2014. As of June 4, 2015:
http://www.rand.org/pubs/working_papers/WR1035.html

"HMA VPN User Arrested After IP Handed Over to the FBI," Hacker10.com, September 28, 2011. As of July 16, 2014:
http://www.hacker10.com/internet-anonymity/
hma-vpn-user-arrested-after-ip-handed-over-to-the-fbi/

Huber, Markus, Martin Mulazzani, and Edgar Weippl, "Tor HTTP Usage and Information Leakage," in *Proceedings of the 11th IFIP TC 6/TC 11 International Conference on Communications and Multimedia Security*, Linz, Austria, May 2010.

"Interview with Bernd Kreuss of TorChat," Free Software Foundation, August 26, 2013. As of June 9, 2015:
https://www.fsf.org/blogs/licensing/interview-with-bernd-kreuss-of-torchat

Joof, Modou S., "'Internet Is Being Used as a Platform for Nefarious and Satanic Purposes,'" Front Page International, July 28, 2013. As of July 16, 2014: http://frontpageinternational.wordpress.com/2013/07/28/ internet-is-being-used-as-platform-for-nefarious-and-satanic-activities/

Kelly, Sanja, Mai Truong, Madeline Earp, Laura Reed, Adrian Shahbaz, and Ashley Greco-Stoner, eds., *Freedom on the Net 2013: A Global Assessment of Internet and Digital Media*, Freedom House, October 3, 2013. As of June 8, 2015: https://freedomhouse.org/sites/default/files/resources/ FOTN%202013_Full%20Report_0.pdf

Kirk, Jeremy, "Hackers Build Private IM to Keep Out the Law," *ComputerWorld. com*, March 28, 2007. As of July 15, 2014: http://www.computerworld.com/s/article/9014675/ Hackers_build_private_IM_to_keep_out_the_law

Kravets, David, "Terrorists Embracing New Android Crypto in Wake of Snowden Revelations," *Ars Technica*, August 1, 2014. As of August 2, 2014: http://arstechnica.com/tech-policy/2014/08/ terrorists-embracing-new-android-crypto-in-wake-of-snowden-revelations/

Li, Bingdong, Esra Erdin, Mehmet Hadi Güneş, George Bebis, and Todd Shipley, "An Analysis of Anonymity Technology Usage," in *Proceedings of the Third International Conference on Traffic Monitoring and Analysis*, Vienna, Austria, April 2011.

"Libyan Uprising One-Year Anniversary: Timeline," *The Telegraph*, February 17, 2012. As of June 5, 2015: http://www.telegraph.co.uk/news/worldnews/africaandindianocean/ libya/9087969/Libyan-uprising-one-year-anniversary-timeline.html

McCoy, Damon, Kevin Bauer, Dirk Grunwald, Tadayoshi Kohno, and Douglas Sicker, "Shining Light in Dark Places: Understanding the Tor Network," in *Proceedings of the 8th International Symposium on Privacy Enhancing Technologies*, Leuven, Belgium, July 2008, pp. 63–76.

Moody, Famiglietti, and Andronico, LLP, *The Tor Project, Inc. and Affiliate: Consolidated Financial Statements and Reports Required for Audits in Accordance with Government Auditing Standards and OMB Circular A-133—December 31, 2013*, Tewksbury, Mass., July 11, 2014. As of June 12, 2015: https://www.torproject.org/about/findoc/2013-TorProject-FinancialStatements.pdf

Norton by Symantec, "The Cybercrime Blackmarket," web page, undated. As of June 30, 2014: http://us.norton.com/cybercrime-blackmarket

O'Carroll, Tanya, "Mobile Technologies Helping Activists and Human Rights Defenders," *Ethical Consumer*, undated. As of June 9, 2015: http://www.ethicalconsumer.org/ethicalreports/mobilesreport/activism.aspx

Patterson, Steven Max, "Mesh Networks and FireChat: How Hong Kong Protestors Are Keeping Communications Alive," *NetworkWorld.com*, October 2, 2014. As of June 8, 2015:
http://www.networkworld.com/article/2691105/opensource-subnet/mesh-networks-and-firechat-how-hong-kong-protestors-are-keeping-communications-alive.html

Pauli, Darren, "Tor Exit Node Mashes Malware into Downloads," *The Register*, October 27, 2014. As of June 9, 2015:
http://www.theregister.co.uk/2014/10/27/tor_exit_node_mashes_malware_into_downloads/

Protalinski, Emil, "Osama bin Laden Didn't Use Encryption: 17 Documents Released," blog post at *ZDNet.com* website, May 3, 2012. As of January 25, 2015:
http://www.zdnet.com/article/osama-bin-laden-didnt-use-encryption-17-documents-released/

Recorded Future, "How Al-Qaeda Uses Encryption Post-Snowden (Part 1)," May 8, 2014a. As of June 5, 2015:
https://www.recordedfuture.com/al-qaeda-encryption-technology-part-1/

———, "How Al-Qaeda Uses Encryption Post-Snowden (Part 2)—New Analysis in Collaboration with ReversingLabs," August 1, 2014b. As of June 5, 2015:
https://www.recordedfuture.com/al-qaeda-encryption-technology-part-2/

Robinson, David, Harlan Yu, and Anne An, *Collateral Freedom: A Snapshot of Chinese Internet Users Circumventing Censorship*, OpenITP, April 2013. As of June 8, 2015:
https://www.teamupturn.com/static/files/CollateralFreedom.pdf

Sandvik, Runa A., "Harvard Student Receives F for Tor Failure While Sending 'Anonymous' Bomb Threat," *Forbes*, December 18, 2013. As of June 30, 2014:
http://www.forbes.com/sites/runasandvik/2013/12/18/harvard-student-receives-f-for-tor-failure-while-sending-anonymous-bomb-threat/

Soghoian, Christopher, "Enforced Community Standards for Research on Users of the Tor Anonymity Network," *Financial Cryptography and Data Security, Lecture Notes in Computer Science*, Vol. 7126, 2012, pp. 146–153.

Thompson, Clive, "How to Keep the NSA Out of Your Computer," *Mother Jones*, September–October 2013. As of June 30, 2014:
http://www.motherjones.com/politics/2013/08/mesh-internet-privacy-nsa-isp

"Timeline: Egypt's Revolution," *Al Jazeera*, February 14, 2011. As of June 5, 2015:
http://www.aljazeera.com/news/middleeast/2011/01/201112515334871490.html

The Tor Project, "Abuse FAQ," web page, undated a. As of June 9, 2015:
https://www.torproject.org/docs/faq-abuse.html.en

———, "Tor: Sponsors," web page, undated b. As of June 9, 2015:
http://www.torproject.org/about/sponsors.html.en

———, "Inception," web page, undated c. As of June 30, 2014:
https://www.torproject.org/about/torusers.html.en

———, "Top-10 Countries by Directly Connecting Users," database, undated d.
As of June 30, 2014:
https://metrics.torproject.org/users.html

———, "We Need Your Good Tor Stories," blog post, August 17, 2011. As of
June 9, 2015:
https://blog.torproject.org/blog/we-need-your-good-tor-stories

———, "Trip Report, October FBI Conference," blog post, December 16, 2012.
As of July 16, 2014:
https://blog.torproject.org/blog/trip-report-october-fbi-conference

———, "Trip Report, Tor Trainings for the Dutch and Belgian Police," blog post,
February 5, 2013a. As of July 16, 2014:
https://blog.torproject.org/blog/trip-report-tor-trainings-dutch-and-belgian-police

———, "How to Handle Millions of New Tor Clients," blog post, September 5,
2013b. As of June 30, 2014:
https://blog.torproject.org/blog/how-to-handle-millions-new-tor-clients

———, "Metrics," database, dated June 30, 2014. As of June 5, 2015:
https://metrics.torproject.org

United States Code, Title 18, Chapter 119, Wire and Electronic Communications
Interception and Interception of Oral Communications (§§ 2510–2522).

United States Code Title 18, Chapter 206, Pen Registers and Trap and Trace
Devices (§§ 3121–3127).

United States Courts, *Wiretap Report 2013*, December 31, 2013. As of March 24,
2015:
http://www.uscourts.gov/statistics-reports/wiretap-report-2013

U.S. Department of State, *Congressional Budget Justification for Fiscal Year 2010*,
May 12, 2009.

———, *Congressional Budget Justification for Fiscal Year 2011, Volume 1:
Department of State Operations*, February 1, 2010.

———, "DRL Internet Freedom Annual Program Statement for Internet Freedom
Technology," web page, April 3, 2013. As of June 12, 2015:
http://www.state.gov/j/drl/p/207061.htm

———, "Bureau of Democracy, Human Rights and Labor Internet Freedom
Annual Program Statement," web page, June 2, 2014. As of June 4, 2015:
http://www.state.gov/j/drl/p/227048.htm

———, *Congressional Budget Justification for Fiscal Year 2016, Appendix 1:
Department of State Diplomatic Engagement*, February 2, 2015.

U.S. Department of State, Bureau of Democracy, Human Rights, and Labor, "DRL Programs," web page, undated. As of June 9, 2015:
http://www.state.gov/j/drl/p/

U.S. House of Representatives, *House Report 112-331—Military Construction and Veterans Affairs and Related Agencies Appropriations Act, 2012*, 2012.

U.S. Senate, *Senate Report 113-81—Department of State, Foreign Operations, and Related Programs Appropriations Bill, 2014*, 2014.

Wickr, "How Wickr Works," undated. As of July 2, 2014:
https://www.mywickr.com/how-wickr-works/

Zetter, Kim, "How the Feds Took Down the Silk Road Drug Wonderland," *Wired.com*, November 18, 2013. As of June 30, 2014:
http://www.wired.com/2013/11/silk-road/